Supplemental Worksheets for Denny Burzynski's

Applied Calculus

for Business, Life, and Social Sciences

Matched Problems

AF817458

Supplemental Worksheets for Denny Burzynski's Applied Calculus for Business, Life, and Social Sciences Matched Problems

Publisher: XYZ Textbooks

Sales: Amy Jacobs, Richard Jones, Rachael Hillman, Bruce Spears, Katherine Hofstra

Editorial Assistants: Rachel Hintz, Katherine Heistand Shields

Proofreader: Judy Barclay

Cover Design: Kyle Schoenberger, Rachel Hintz

Cover Image: © Matthew Bowden

Printed in the United States of America

ISBN-13: 978-1936368-94-5 / ISBN-10: 1-936368-94-3

For product information and technology assistance, contact us at
XYZ Textbooks, 1-877-745-3499

For permission to use material from this text or product,
e-mail: **info@mathtv.com**

XYZ Textbooks
1339 Marsh Street
San Luis Obispo, CA 93401
USA

For your course and learning solutions, visit **www.xyztextbooks.com**

Ticket to Success

Dear Student,

These worksheets may be your ticket to success in this course. If you are consistent about using these worksheets you will never get behind in the course, and you will always be prepared for what is presented in class. If you successfully complete every one of these worksheets, you will have successfully worked every type of problem presented in the course.

Start a worksheet before you go to class. After you have finished a problem, check your answer. If you have made a mistake, find it and correct your work. If you get stuck on a problem, watch one of the videos associated with the problem. There are videos that match each of the problems on your worksheet.

If you have decided that these worksheets are a good idea for you, then make a commitment to work on them every day. Don't just try a few and see how it goes. Make a commitment and stick to it. By doing so you are saying that you want to be successful in the course and are willing to go to any length to achieve that success.

Pat McKeague
Fall 2014

xyztextbooks

MathTV.com

Innovative, Quality Educational Materials
reasonable prices

Matched Problems
Applied Calculus
Section 1.1

Name ______________________________

Date ______________________________

Directions Each problem below is similar to the example with the same number in your textbook. After reading through an example in your textbook, or watching one of the videos of that example on MathTV.com, try the matched problem to check your progress in this section. You can use the QR code at the right to view the videos for this section on your smart phone or tablet. If your smart device does not have a camera, but can access the internet, go to m.mathtv.com to view the videos for this worksheet.

1. Construct a table and graph for $y = 8x, 0 \le x \le 40$.

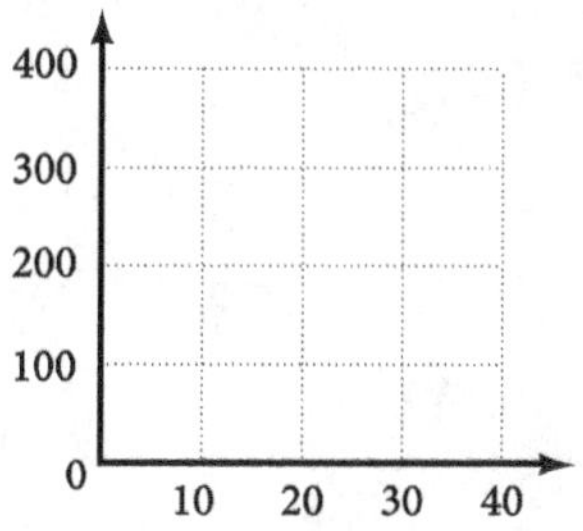

Function Rule

x	$y = 8x$	y
0		
10		
20		
30		
40		

2. State the domain and range for $y = 8x, 0 \le x \le 40$.

3. Kendra is tossing a softball into the air so that the distance h the ball is above her hand t seconds after she begins the toss is given by $h = 48t - 16t^2$, $0 \le t \le 3$. Construct a table and line graph for this function.

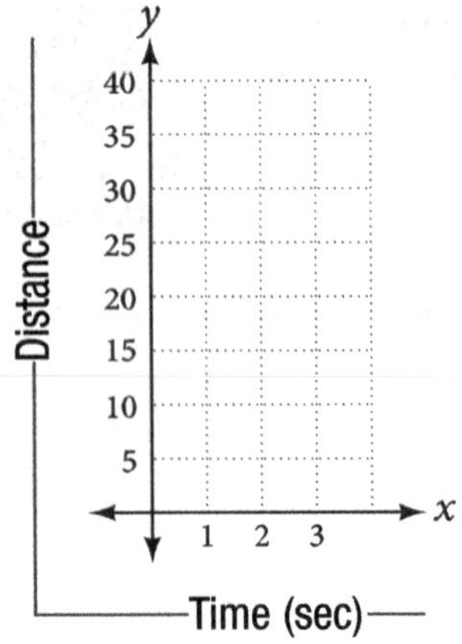

Time (sec) t	*Function Rule* $h = 48t - 16t^2$	*Distance* h
0		
1		
1.5		
2		
3		

4. Graph the following set of ordered pairs and then determine whether this relation is a function:

$\{(1, 5), (2, 4), (4, 8), (2, 9), (5, 10), (6, 12)\}$

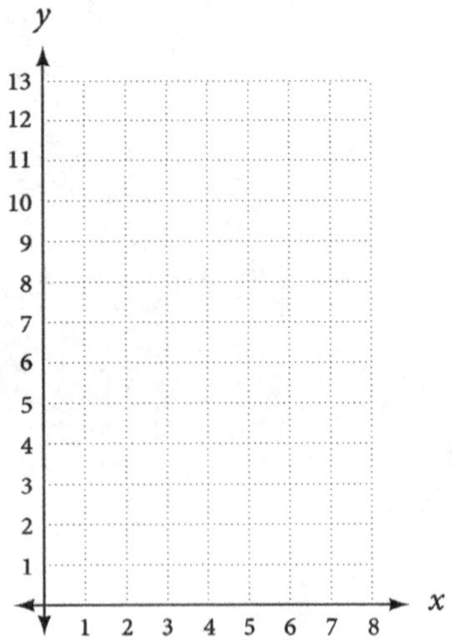

5. Use the equation $x = y^2 - 4$ to fill in the following table then use the table to sketch the graph. Is this graph a function?

y	*Function Rule* $x = y^2 - 4$	x
−3		
−2		
−1		
0		
1		
2		
3		

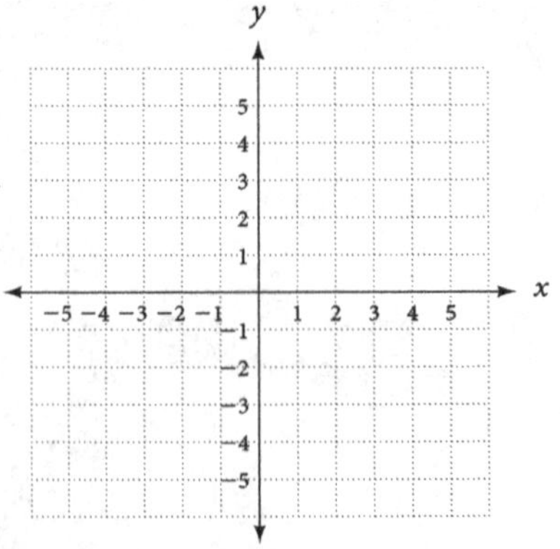

6. Graph $y = \dfrac{2}{x}$. Is this the graph of a function?

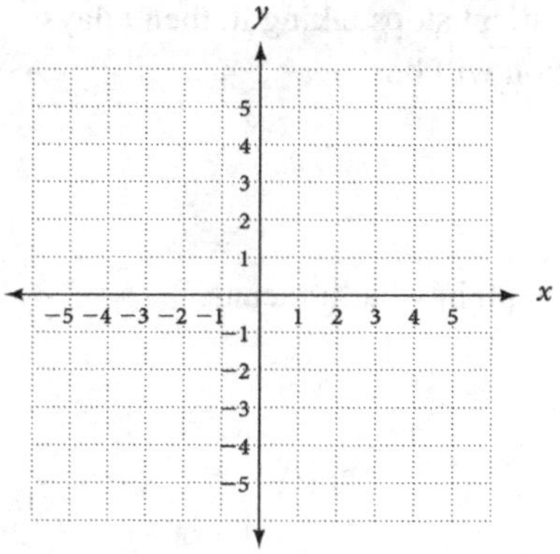

7. Graph $y = |x| - 3$. Is this the graph of a function?

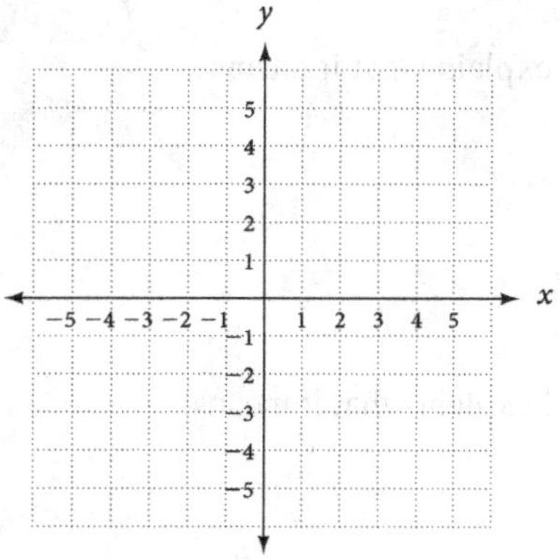

8. If $f(x) = 8x$, find each of the following:

a. $f(0)$

b. $f(5)$

c. $f(10.5)$

9. If $f(x) = 2x + 1$ and $g(x) = x^2 - 3$, find the following:

a. $f(5)$

b. $g(5)$

c. $f(-2)$

d. $g(-2)$

e. $f(a)$

f. $g(a)$

SECTION 1.1

10. When Lorena runs a mile in t minutes, her average speed in feet per seconds is given by $s(t)=\frac{88}{t},\ t>0$.

a. Find $s(8)$ and explain what it means.

b. Find $s(11)$ and explain what it means.

11. A medication has a half-life of 5 days. If the concentration of the medication in a patient's system is 80 ng/mL and the patient stops taking it, then t days later the concentration will be:

$$C(t)=80\left(\frac{1}{2}\right)^{t/5}$$

a. Find $C(5)$ and explain what it means.

b. Find $C(10)$ and explain what it means.

12. The following formulas give the circumference and area of a circle with a radius of r. Use the formulas to find the circumference and area of a circular plate if the radius is 5 inches.

$C(r)=2\pi r$ $\qquad$ $A(r)=\pi r^2$

Math TV.com

Matched Problems
Applied Calculus
Section 1.2

Name ______________________________

Date ______________________________

Directions Each problem below is similar to the example with the same number in your textbook. After reading through an example in your textbook, or watching one of the videos of that example on MathTV.com, try the matched problem to check your progress in this section. You can use the QR code at the right to view the videos for this section on your smart phone or tablet. If your smart device does not have a camera, but can access the internet, go to m.mathtv.com to view the videos for this worksheet.

1. If $f(x) = 3x + 2$, $g(x) = 3x^2 - 10x - 8$, and $h(x) = x - 4$, find formulas for the following:

a. $f + g$

b. fh

c. fg

d. $\frac{g}{f}$

2. If $f(x) = 4x - 3$, $g(x) = 4x^2 - 7x + 3$, and $h(x) = x - 1$, find the following:

a. $(f + g)(3)$

b. $(fh)(-1)$

c. $(fg)(0)$

d. $\left(\frac{g}{f}\right)(10)$

SECTION 1.2

3. Repeat example 3 if she wants the weekly revenue to be \$4,200.

4. If $f(x)=x-4$ and $g(x)=x^2+3x$, find the following:

a. $(f\circ g)(x)$

b. $(g\circ f)(x)$

5. Repeat example 5 and find the number of cans sold and recycled when the temperature is 37°C.

Matched Problems
Applied Calculus
Section 1.3

Name ____________________

Date ____________________

Directions Each problem below is similar to the example with the same number in your textbook. After reading through an example in your textbook, or watching one of the videos of that example on MathTV.com, try the matched problem to check your progress in this section. You can use the QR code at the right to view the videos for this section on your smart phone or tablet. If your smart device does not have a camera, but can access the internet, go to m.mathtv.com to view the videos for this worksheet.

1. Graph the line $y = 3x - 5$ and then find its slope.

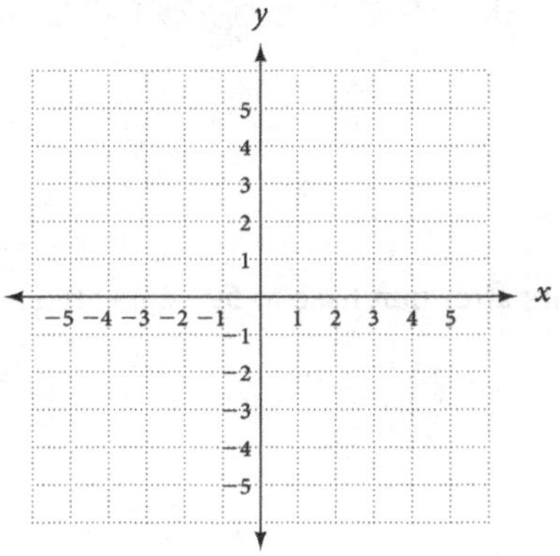

2. Find the slope of the line which passes through $(3, 4)$ and $(1, -2)$.

3. Find the slope of the line which passes through $(2, -3)$ and $(-1, -3)$.

4. Using the chart from example 4 of your textbook, find the slope of the line connecting the point $(1985, 1.25)$ with the last point $(2005, 2.93)$. Explain what your result represents.

5. Use the equation from example 5 in your textbook to find the average rate of change from 1 second to $\frac{5}{4}$ seconds.

6. Referring to example 6 in your book, find the average rate of change from 10 years after purchase to 50 years after purchase.

7. Use the formula $S(r) = 4\pi r^2 \quad r \geq 0$ to find the average rate of change in the surface area of the balloon as the radius increases from 4 inches to 6 inches.

8. If $f(x)=3x-9$, find $\dfrac{f(b)-f(a)}{b-a}$ and simplify.

9. If $f(x)=x^2-16$, find $\dfrac{f(x_2)-f(x_1)}{x_2-x_1}$ and simplify.

10. Find the equation of the line with slope $\dfrac{2}{3}$ and y-intercept 1.

11. Give the slope and y-intercept for the line $4x-5y=7$.

12. Give the slope and y-intercept for the equation $-3x+2y=-6$, then use the slope and y-intercept to graph the line.

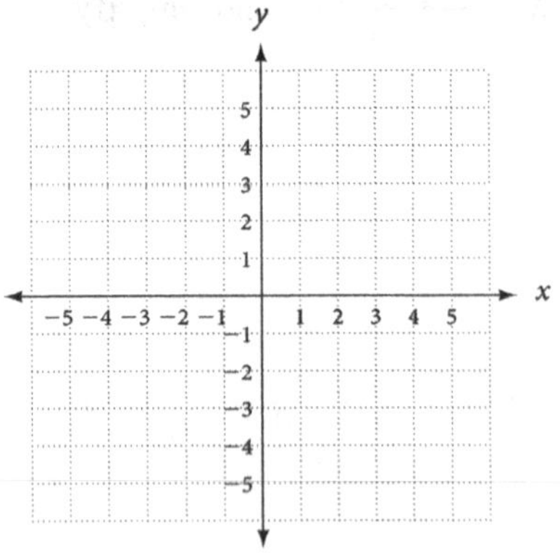

13. Find the equation of the line with slope 3 that contains the point $(-1,\ 2)$. Write the answer in slope-intercept form.

14. Find the equation of the line that contains the points $(2, 5)$ and $(6, -3)$. Write the answer in slope-intercept form.

15. Graph the function defined by:

$$f(x)=\begin{cases} x+2 & \text{if } x \le 1 \\ 5 & \text{if } x > 1 \end{cases}$$

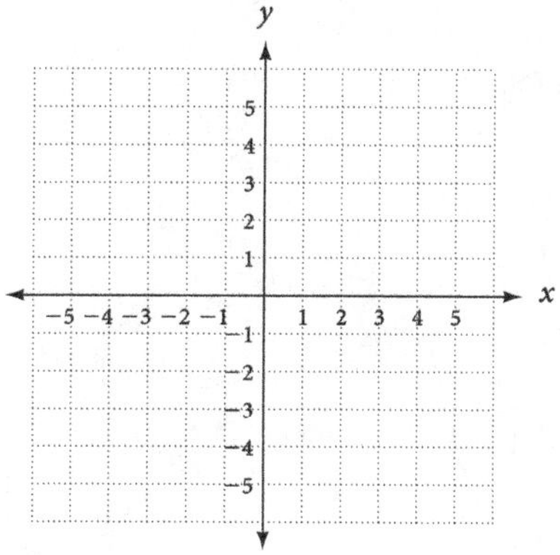

16. A fraternity is hosting a charity fundraiser and their profit P is given by the function

$$P(x)=\begin{cases} 10x-200 & 0\le x\le 150 \\ 10x-350 & 150<x\le 350 \end{cases}$$

where x is the number of guests who attend. Use this function to find the following:

SECTION 1.3

a. Domain **b.** $P(140)$ **c.** $P(150)$ **d.** $P(160)$

Matched Problems
Applied Calculus
Section 1.4

Name ______________________

Date ______________________

SECTION 1.4

Directions Each problem below is similar to the example with the same number in your textbook. After reading through an example in your textbook, or watching one of the videos of that example on MathTV.com, try the matched problem to check your progress in this section. You can use the QR code at the right to view the videos for this section on your smart phone or tablet. If your smart device does not have a camera, but can access the internet, go to m.mathtv.com to view the videos for this worksheet.

1. Find the limit as x approaches 2, of the function

$$f(x)=\frac{x^2-4}{x-2}$$

2. Find each limit for the function below:

$$P(x)=\begin{cases} 10x-200 & 0\le x\le 150 \\ 10x-350 & 150<x\le 350 \end{cases}$$

a. $\lim_{x\to 140} P(x)$

b. $\lim_{x\to 150} P(x)$

c. $\lim_{x\to 160} P(x)$

3. If $f(x)=-3$, find $\lim_{x\to 7} f(x)$.

4. If $f(x)=2x^2-x+3$, find $\lim_{x\to 2} f(x)$.

5. If $f(x)=\frac{x+5}{x^2-3}$, find $\lim_{x\to -2} f(x)$.

6. Find $\lim_{x\to -1}\left(x+5-\frac{4}{x}\right)$.

SECTION 1.4

7. Find $\lim_{x\to 1}\left[(4x-1)(x^2+2)\right]$.

8. Find $\lim_{x\to -2}\dfrac{5x-4}{x-5}$.

9. Find $\lim_{x\to 1/2}(4x-1)^3$.

10. Find $\lim_{x\to 4}\sqrt{x^3}$.

11. If $f(x)=\dfrac{1}{(x-2)^2}$, find:

a. $\lim_{x\to 1} f(x)$

b. $\lim_{x\to 2} f(x)$

c. $\lim_{x\to 3} f(x)$

12. Find $\lim_{x\to 2}\dfrac{x^2+x-5}{x-2}$.

13. Find $\lim_{x \to -1} \dfrac{x^2 - x - 2}{x + 1}$.

14. Find $\lim_{x \to 9} \dfrac{\sqrt{x} - 3}{x - 9}$.

15. Find $\lim_{x \to \infty} \dfrac{5x^2 - 2x + 2}{7x^2 + 5x - 7}$.

16. Find $\lim_{x \to \infty} \dfrac{2x^2 - 5}{3x - 2}$.

17. Find $\lim_{x \to \infty} \dfrac{3x - 2}{2x^2 - 5}$.

18. Find each of the following limits:

a. $\lim_{x \to \infty} \dfrac{3x - 5}{x^2 + 1}$.

b. $\lim_{x \to \infty} \dfrac{3x - 5}{x + 1}$.

c. $\lim_{x \to \infty} \dfrac{3x^2 - 5}{x + 1}$.

19. Repeat example 19 if:

$$W(x) = \frac{160(x+7)}{x+5} \text{ if } x \geq 0$$

MathTV.com

Matched Problems
Applied Calculus
Section 1.5

Name ______________________

Date ______________________

SECTION 1.5

Directions Each problem below is similar to the example with the same number in your textbook. After reading through an example in your textbook, or watching one of the videos of that example on MathTV.com, try the matched problem to check your progress in this section. You can use the QR code at the right to view the videos for this section on your smart phone or tablet. If your smart device does not have a camera, but can access the internet, go to m.mathtv.com to view the videos for this worksheet.

1. Discuss the continuity of the function at $x=2$:

a. $f(x)=\dfrac{x^2-3x-2}{x-2}$

b. $g(x)=\begin{cases} x^2-3x-2 & x\neq 2 \\ 3 & x=2 \end{cases}$

c. $h(x)=\begin{cases} \dfrac{x^2-3x-2}{x-2} & x\neq 2 \\ 5 & x=2 \end{cases}$

2. Discuss the continuity of the function at $x=2$:

a. $f(x)=\begin{cases} 2x-1 & x\geq 2 \\ x+1 & x<2 \end{cases}$

b. $g(x)=\begin{cases} 2x-1 & x\geq 2 \\ x+3 & x<2 \end{cases}$

c. $h(x)=\begin{cases} 2x-1 & x>2 \\ x+1 & x<2 \end{cases}$

3. Determine the continuity of the function at each of the given points:

$P(x)=\begin{cases} 10x-200 & 0\leq x\leq 150 \\ 10x-350 & 150<x\leq 350 \end{cases}$

a. $x=140$

b. $x=150$

c. $x=160$

4. Is the function described by the graph below continuous at $x = a$?

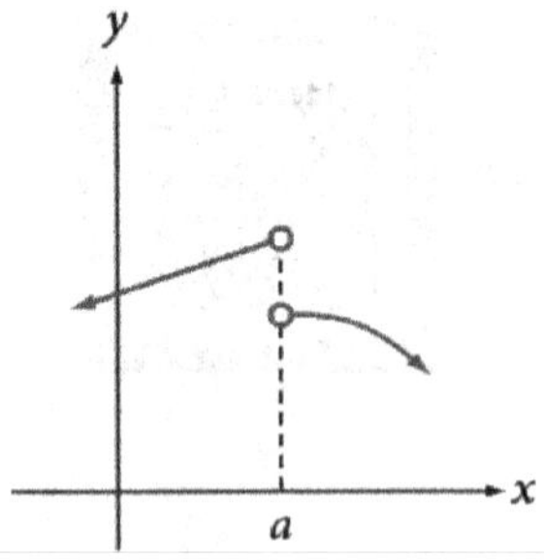

5. Use interval notation to show where each function is continuous:

a. $f(x) = x^2 - 4$

b. $g(x) = \dfrac{x+1}{x^2 - x - 20}$

6. Refer to the graph at the opening of this section. Is this an idealized or realistic model of profit? If idealized, discuss how the actual graph of profit would differ from the graph given.

Matched Problems
Applied Calculus
Section 1.6

Name ______________________

Date ______________________

SECTION 1.6

Directions Each problem below is similar to the example with the same number in your textbook. After reading through an example in your textbook, or watching one of the videos of that example on MathTV.com, try the matched problem to check your progress in this section. You can use the QR code at the right to view the videos for this section on your smart phone or tablet. If your smart device does not have a camera, but can access the internet, go to m.mathtv.com to view the videos for this worksheet.

1. For the function $f(x)=2\sqrt{3x}+2$, find the average rate of change with respect to x on the interval $[3, 27]$.

2. For the function $P(x)=-x^2+13x-28$, find the average rate of change with respect to x on the interval $[3, 4]$.

3. For the function $f(x)=7x+3$, find the difference quotient $\frac{f(x+h)-f(x)}{h}$.

4. For the function $f(x)=-x^2+13x-22$, find and simplify the difference quotient $\frac{f(x+h)-f(x)}{h}$.

SECTION 1.6

5. Find the instantaneous rate of change of the function $f(x)=7x^2+2x-4$ at

a. $x=1$ **b.** $x=2$ **c.** $x=-4$

6. Repeat example 6 using the function $C(x)=-4x^2+250x$.

7. Repeat example 7 using the function $s(t)=64t-16t^2$.

MathTV.com

Matched Problems

Applied Calculus
Section 2.1

Name ____________________

Date ____________________

Directions Each problem below is similar to the example with the same number in your textbook. After reading through an example in your textbook, or watching one of the videos of that example on MathTV.com, try the matched problem to check your progress in this section. You can use the QR code at the right to view the videos for this section on your smart phone, or tablet. If your smart device does not have a camera, but can access the internet, go to m.mathtv.com to view the videos for this worksheet.

1. Find the derivative of $f(x) = 6x + 9$ and interpret its meaning.

2. For the function $f(x) = x^2 - 8x + 4$,

a. Use the definition of the derivative to find and interpret $f'(x)$.

b. Find $f'(3)$.

3. Interpret the meaning of each:

a. If $f(x) = 5, f'(3) = 0$

b. If $f(x) = 12, f'(x) = 0$.

4. Find $f'(x)$ for each function.

a. $f(x) = x^7$

b. $f(x) = x^{10}$

c. $f(x) = x$

SECTION 2.1

5. If $f(x) = \sqrt[4]{x^3}$, find

a. $f(16)$

b. $f'(16)$

6. If $f(x) = \frac{1}{x^2}$, find

a. $f'(x)$

b. The equation of the tangent line at $(1, 1)$

7. If $f(x) = \frac{6}{x^7}$, find $f'(x)$.

8. If $y = 7x$, find $\frac{dy}{dx}$.

9. Find the derivative of each function.

a. $y = 9x$

b. $f(x) = -12x$

10. If $f(x) = 4x^3 - 7x^2 - 8x + 1$, find $f'(x)$.

11. If $f(x) = x^2 - x - 20$, find

a. the value of x for which $f(x) = 0$.

b. the value of x for which $f'(x) = 0$.

12. If $y = 8\sqrt[3]{x}$, find $\left.\frac{dy}{dx}\right|_{x=27}$.

13. If $y = 12\sqrt[4]{x}$, find and interpret $\left.\frac{dy}{dx}\right|_{x=16}$.

14. Repeat parts a and b from Example 14 from the textbook with the function, $N(t) = 100t^2 - t^3$.

Matched Problems
Applied Calculus
Section 2.2

Name ______________________

Date ______________________

Directions Each problem below is similar to the example with the same number in your textbook. After reading through an example in your textbook, or watching one of the videos of that example on MathTV.com, try the matched problem to check your progress in this section. You can use the QR code at the right to view the videos for this section on your smart phone, or tablet. If your smart device does not have a camera, but can access the internet, go to m.mathtv.com to view the videos for this worksheet.

1. Find the derivative of $f(x) = (2x^2 - 1)(x^2 + 3)$ by
   a. multiplying the terms together, then differentiation the product, and
   b. using the product rule.

   Compare your answers. They should be the same.

2. For the function $g(t) = (t^2 - 9)(t^2 - 1)$, find
   a. the values of t for which $g(t) = 0$.
   b. the values of t for which $g'(t) = 0$.

3. For the function $f(x) = \dfrac{x^3 + 5x + 4}{x^2 - 3}$, find
   a. $f'(x)$
   b. $f'(0)$

4. Suppose that Analyst C uses the revenue model to predict the revenue of the Polynomial car company t years from now. For her revenue function, she computes $R_C(5) = 42$ and $R'_C(5) = 14$. Analyst B uses the revenue model $R_B(t) = 4$ to predict the revenue of the same Polynomial car company also t years from now. For his revenue function, he computes $R_D(5) = 6$ and $R'_D(5) = 2$.

a. What information is contained in the quotient

$$\frac{R_C(5)}{R_D(5)} = \frac{42}{6} = 7$$

SECTION 2.2

b. What information is contained in the quotient

$$\frac{R'_C(5)}{R'_D(5)} = \frac{14}{2} = 7$$

c. What information is contained in the quotient

$$\left[\frac{R_C(t)}{R_D(t)}\right]' = 3 \text{ when } t = 5$$

5. Repeat Example 5 from the textbook using the function

$$P(t) = \frac{600 + 100t - 10t^2}{t^2 + 6}.$$

6. Find the equation of the line tangent to the graph of

$$f(x) = \frac{5x + 4}{x - 2}$$

when $x = 1$.

Matched Problems
Applied Calculus
Section 2.3

Name ______________________

Date ______________________

Directions Each problem below is similar to the example with the same number in your textbook. After reading through an example in your textbook, or watching one of the videos of that example on MathTV.com, try the matched problem to check your progress in this section. You can use the QR code at the right to view the videos for this section on your smart phone, or tablet. If your smart device does not have a camera, but can access the internet, go to m.mathtv.com to view the videos for this worksheet.

1. **a.** Find $f'(x), f''(x), f'''(x)$, and $f^{(4)}(x)$, for

$$f(x) = 2x^5 - 3x^4 + 7x^2 + 5x - 12$$

b. Find $f'(x), f''(x), f'''(x)$, and $f^{(4)}(x)$, for

$$f(x) = 8x^3 - 2x + 1$$

2. For the function $f(x) = \frac{1}{3}x^6 - 3x^4$, find

a. $f(2)$

b. $f'(2)$

c. $f''(2)$

3. For the function $g(t) = 3t^3 - 12t$, find the value(s) of t for which

a. $g(t) = 0$

b. $g'(t) = 0$

c. $g''(t) = 0$

4. For the function $y = \frac{2}{x^2}$, find y', y'', y'''.

5. Suppose the function $W(x)$ represents the amount of money (in millions of dollars) private contributors donate annually x years after the year 2000 to researchers searching for a cause and cure of colitis. Interpret the information provided by $W(8) = 42$, $W'(8) > 0$, and $W''(8) < 0$.

6. Suppose the function $C(t)$, where t represents the number of years from January 1, 2000, approximates the health care cost to a city to care for uninsured citizens. Interpret the information provided by

$$C'(13) = -0.1 \text{ and } C''(13) < 0$$

7. Suppose the function $C(t)$, where t represents the number of months from January 1, 2012, approximates the energy cost to a manufacturing company. Interpret the information provided by

$$C'(13) = -0.6 \text{ and } C''(13) > 0.$$

Matched Problems
Applied Calculus
Section 2.4

Name ____________________

Date ____________________

Directions Each problem below is similar to the example with the same number in your textbook. After reading through an example in your textbook, or watching one of the videos of that example on MathTV.com, try the matched problem to check your progress in this section. You can use the QR code at the right to view the videos for this section on your smart phone, or tablet. If your smart device does not have a camera, but can access the internet, go to m.mathtv.com to view the videos for this worksheet.

1. Suppose that $y = u^3 + 4$ and $u = 5x - 1$

a. Find $\frac{dy}{dx}$.

b. Evaluate and interpret $\frac{dy}{dx}$ when $x = 2$.

2. The demand, d, for a product depends on the price, p, according to the function, $d(p) = 64p^{-2}$, where d is the numbers of units and p is in dollars. The cost, c, per unit sold depends on the number of units produced according to the function, $c(d) = 20 + 10d^{-\frac{1}{3}}$, where c is in dollars.

a. Write a function that shows how the cost of producing a unit changes with the sales price.

b. How does the cost of production change as the price increases from \$27 to \$28 per unit?

3. Find the derivative of each function

a. $f(x) = (5x^2 + 6x - 8)^4$

b. $f(x) = \dfrac{5}{(4 - 7x)^3}$

c. $f(x) = \sqrt[6]{5x - 2}$

4. Find the derivative of the function

$$f(x) = \frac{(4x - 5)^3}{5x + 7} \quad \text{at } x = -1$$

5. Find the derivative of the function $f(x) = x^2(x + 5)^{1/3}$.

6. Find $f'(x)$ for the function $f(x) = \left(\frac{3x - 4}{5x + 2}\right)^6$

Matched Problems
Applied Calculus
Section 2.5

Name ____________________

Date ____________________

Directions Each problem below is similar to the example with the same number in your textbook. After reading through an example in your textbook, or watching one of the videos of that example on MathTV.com, try the matched problem to check your progress in this section. You can use the QR code at the right to view the videos for this section on your smart phone, or tablet. If your smart device does not have a camera, but can access the internet, go to m.mathtv.com to view the videos for this worksheet.

1. **a.** Find an expression for y' for $4x^3 + 3y^4 + y = 15$.
b. Find y' at (3, 1).

2. Find y' for $x^4y^4 - 4x^2 + 5y = 30$

3. Suppose both y and x are differentiable functions of t and that the relationship between y and x is expressed by the equation $2x^4 + 3y^2 = 59$. Find and interpret $\frac{dy}{dt}$ when $\frac{dx}{dt} = 5$ and $x = 2$ and $y = 3$.

4. Repeat Example 4 in textbook when the rate is 2.5 feet per hour and the radius of the spill is 650 feet.

5. Repeat Example 5 in the textbook using the function

$$x = \frac{20{,}000}{\sqrt[3]{p^2-5}} + 300.$$

Matched Problems
Applied Calculus
Section 3.1

Name ____________________

Date ____________________

Directions Each problem below is similar to the example with the same number in your textbook. After reading through an example in your textbook, or watching one of the videos of that example on MathTV.com, try the matched problem to check your progress in this section. You can use the QR code at the right to view the videos for this section on your smart phone, or tablet. If your smart device does not have a camera, but can access the internet, go to m.mathtv.com to view the videos for this worksheet.

1. Use the graph of the function $f(x) = x^3 - x^2 - 8x - 17$ to estimate the intervals upon which $f(x)$ is increasing and upon which it is decreasing.

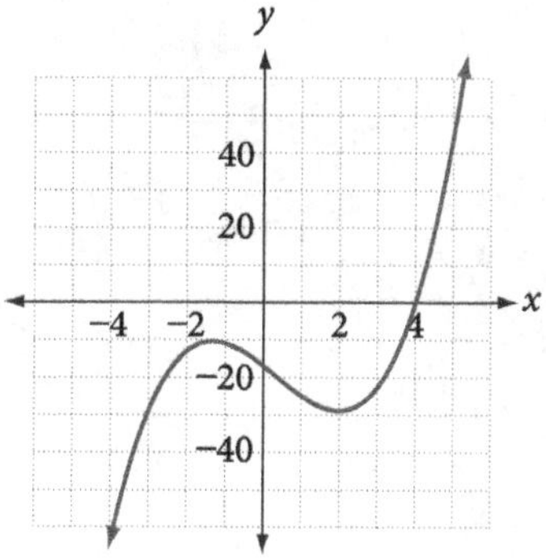

2. Construct a sign chart to help you find the interval upon which the function

$$f(x) = 3x^2 - 15x$$

is increasing and the interval upon which it is decreasing.

3. Discuss the first derivative of $g(x) = 7$ and its relationship to its graph.

4. Use the first derivative test to find the relative extrema of the function

$$f(x) = \frac{1}{2}x^4 - 36x^2$$

Summarize your results in a summary table.

5. Construct a sign chart for the graph pictured and use it to construct a summary table for the function

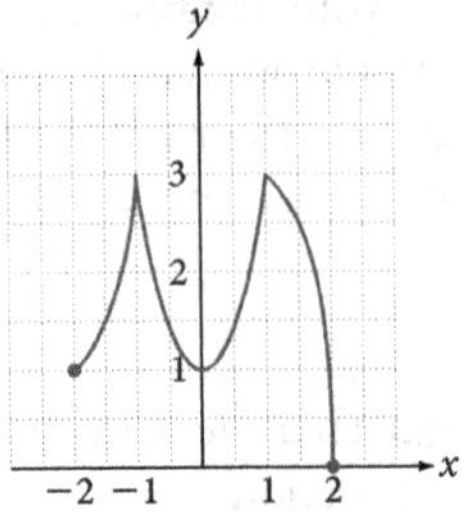

6. Repeat example 6 using the function

$$C(x) = \frac{10t}{5t^2 + 20}$$

7. Find, if they exist, the absolute maximum and an minimum of the function

$$f(x) = (x - 5)^{1/3} + 3$$

on the interval [0, 8].

8. Find, if they exist, all relative and absolute extrema of the function

$$f(x) = \frac{1}{x - 7}$$

Matched Problems
Applied Calculus
Section 3.2

Name ____________________

Date ____________________

Directions Each problem below is similar to the example with the same number in your textbook. After reading through an example in your textbook, or watching one of the videos of that example on MathTV.com, try the matched problem to check your progress in this section. You can use the QR code at the right to view the videos for this section on your smart phone, or tablet. If your smart device does not have a camera, but can access the internet, go to m.mathtv.com to view the videos for this worksheet.

1. An efficiency study conducted on the 12-hour night shift of a manufacturer showed that the number of units of a product produced by a typical employee t hours after 4:00 PM could be modeled by the function

$$f(t) = -2t^3 + 24t^2 \quad 0 \leq t \leq 12$$

In context of the problem, describe the behavior of this function.

2. Describe the behavior of the function

$$f(x) = x^{2/7} + 1$$

3. Describe the behavior of the function

$$f(x) = \frac{1}{3}x^3 - 6x^2 + 4$$

4. Describe the behavior of the function

$$f(x) = \frac{7x + 3}{7x - 3}$$

5. Repeat example 5 using the function

$$R(t) = -\frac{1}{4}x^4 + 450x^2$$

6. Use the second derivative test to find all the relative extrema of the function

$$f(x) = 3x^3 + x^2 - 7x + 1$$

Matched Problems
Applied Calculus
Section 3.3

Name ____________________

Date ____________________

Directions Each problem below is similar to the example with the same number in your textbook. After reading through an example in your textbook, or watching one of the videos of that example on MathTV.com, try the matched problem to check your progress in this section. You can use the QR code at the right to view the videos for this section on your smart phone, or tablet. If your smart device does not have a camera, but can access the internet, go to m.mathtv.com to view the videos for this worksheet.

1. A rectangular area along the bank of a straight section of a river is to be enclosed with 280 feet of fencing. Only three sides need to be fenced since the river will bound the other side. What are the dimensions that will produce the greatest enclosed area?

2. The concentration $C(t)$ (in mg/cm^3) of a drug in a patient's bloodstream t hours after the drug is administered is approximated by the function

$$C(t) = \frac{0.03t}{t^2 + 14t + 49}$$

Find the maximum concentration and the time at which it occurs.

3. A company can sell 10 reels of wire per week at a price of \$150 each. The manager estimates that for each \$5 reduction in price, she can sell one more reel per week. The reels cost the company \$80 each. Find the best price and the quantity that will maximize the store's profit. What is that profit?

4. A rectangular box with an open top is to be made from a square piece of cardboard that is 8 inches on each side by removing small squares of equal size from the corners and folding up the remaining flaps. What should be the size of the squares cut from the comers so that the box will have maximum volume?

5. Repeat example 5 using the function

$$P(t) = \frac{(t-6)^2}{t+1}$$

Matched Problems
Applied Calculus
Section 3.4

Name ____________________

Date ____________________

Directions Each problem below is similar to the example with the same number in your textbook. After reading through an example in your textbook, or watching one of the videos of that example on MathTV.com, try the matched problem to check your progress in this section. You can use the QR code at the right to view the videos for this section on your smart phone, or tablet. If your smart device does not have a camera, but can access the internet, go to m.mathtv.com to view the videos for this worksheet.

1. The revenue realized by a company on the sale of x units of one of its products is approximated by

$$R(x) = \frac{1}{3}x^3 - \frac{7}{2}x^2 - 300x + 200$$

Compute and interpret both $R(85)$ and $R'(85)$.

2. Data indicates that the cost to a company of shipping x units of product is approximately

$$C(x) = 0.0001x^3 - 0.002x^2 + 0.03x$$

Compute and interpret both $C(350)$ and $C'(350)$.

3. The cost to a company to produce x units of a product is given by the cost function

$$C(x) = 0.01x^3 - 0.4x^2 + 15x + 70$$

The revenue the company expects to realize from the sale of x units of the product is given by the revenue function

$$R(x) = -0.14x^2 + 65x$$

If $P(x)$ represents the profit realized on the sale of x units of the product, compute and interpret both $P(40)$ and $P'(40)$.

4. A company estimates that it can sell 60 units each month of its product if it prices each unit at $200. However, its monthly number of sales will increase by 5 units for each $20 decrease in price. The company has fixed costs of $1000. The cost to make each unit is $40. Find the level of production that maximizes the company's profit if the company must produce and sell between and including 40 and 95 units each month.

5. A company anticipates selling 1,000 units of its product at a uniform rate over the next year. Each time the company places an order for x units, it is charged a flat fee of \$240. Carrying costs are \$300 per unit per year. How many times should the company reorder each year and what should be the lot size of each order to minimize inventory costs.

6. Using the information from problem 5, use the EOQ to get the lot size that will minimize inventory cost.

7. A company estimates that the weekly sales q of its product is related to the product's price p by the function

$$q = \frac{4{,}800}{\sqrt[3]{p^2}}$$

where p is in dollars. Currently, each unit of the product sells for \$64. Determine the point elasticity of demand of this product.

8. **a.** In Matched Problem 7, we computed ε to be 0.67. The number of units sold went from 300 to 298 as the price increased 1% from \$64 to \$64.64. Is the demand elastic, unit elastic, or inelastic? What is the revenue when the price of each unit was \$64? What is the revenue with a price decrease of 1%?

   b. Suppose a particular state is considering adding a 1% tax on its college and university tuitions. State economists advise that doing so produces an ε value of 1.3. Should or should not the state adopt a 1% increase in tuition?

Matched Problems
Applied Calculus
Section 4.1

Name ____________________

Date ____________________

Directions Each problem below is similar to the example with the same number in your textbook. After reading through an example in your textbook, or watching one of the videos of that example on MathTV.com, try the matched problem to check your progress in this section. You can use the QR code at the right to view the videos for this section on your smart phone, or tablet. If your smart device does not have a camera, but can access the internet, go to m.mathtv.com to view the videos for this worksheet.

1. a. Construct both a table and graph of the functions $f(x) = 3^x$ and $g(x) = \left(\frac{1}{3}\right)^x$ from $x = -3$ to $x = 3$.
   b. Construct both a table and graph of the functions
   $$f(x) = \left(\frac{3}{2}\right)^x \quad \text{and} \quad g(x) = \left(\frac{2}{3}\right)^x$$
   from $x = -3$ to $x = 3$.

2. On the same coordinate system, construct a graph of the functions of $f(x) = e^{1x}$, $f(x) = e^{2x}$, and $f(x) = e^{0.2x}$ from $x = -2$ to $x = 2$.

SECTION 4.1

3. On the same coordinate system, construct a graph of the functions of $f(x) = e^{-1x}$, $f(x) = e^{-2x}$, and $f(x) = e^{-0.2x}$ from $x = -2$ to $x = 2$.

4. Using the function $A(t) = 7e^{0.03169t}$, where t is in minutes, how many bacteria could be predicted to be on the counter 3 hours later?

5. Using the function from example 5, $N(P) = N_1 e^{-0.11(p-1)}$, if in a discussion group of 10 people, the first-ranked person participated 21 times, how many times did the fourth-ranked person participate?

6. Determine the amount of money accumulated after 6 years if $4,000 is invested in an account that pays 3.85% interest compounded weekly.

7. Repeat example 7 using the function

$$N(t) = \frac{150{,}000}{1 + 5.9e^{-0.78(t-2010)}}$$

Matched Problems
Applied Calculus
Section 4.2

Name ______________________

Date ______________________

Directions Each problem below is similar to the example with the same number in your textbook. After reading through an example in your textbook, or watching one of the videos of that example on MathTV.com, try the matched problem to check your progress in this section. You can use the QR code at the right to view the videos for this section on your smart phone, or tablet. If your smart device does not have a camera, but can access the internet, go to m.mathtv.com to view the videos for this worksheet.

1. Convert each exponential function to the corresponding logarithmic function.
- **a.** $y = e^{10x}$
- **b.** $z = e^{6x-6}$
- **c.** $450 = e^{-0.05x}$

2. Convert each logarithmic function to the corresponding exponential function.
- **a.** $10 = \ln(5x)$
- **b.** $x + 1 = \ln(y + 4)$
- **c.** $-0.005 = \ln(x - 10)$

3. Evaluate $\ln e^x$.

4. Use the properties of natural logarithms to expand each logarithmic expression.
- **a.** $\ln(5x)$
- **b.** $\ln\left(\dfrac{3x}{x+9}\right)$
- **c.** $\ln x^4$
- **d.** $150 \ln(x^3y^7)$

5. Use the properties of the natural logarithm to write each logarithmic expression as an expression with a single logarithm.

a. $9 \ln b + \ln a$

b. $\ln d + \ln g - \ln h$

6. To four decimal places, approximate the solution to the equation

$$e^{6x+3} = 14$$

SECTION 4.2

7. The exponential function $N(x) = 2e^{0.1115x}$ estimates the number of weblogs (blogs) created each day between March, 2003, and February, 2005. The variable N represents the number of thousands of blogs created per day x months after March, 2003. Estimate the number of months after March, 2003, when 10,000 blogs per day were being created. In this case, $N = 10$ (for 10,000).

Source: David Silfry's Musings at Silfry's Alerts
http://www.sifry.com/alerts/archives/000298.html

8. XYZ Textbooks bought a piece of capital equipment for \$265,000. It depreciates according to the function $A = 265{,}000e^{-0.27t}$, where t represents the number of years since its purchase. The company will replace the piece of equipment when the piece has fallen to 20% of its original value. To the nearest year, how many years will XYZ Textbooks keep the piece of equipment?

Matched Problems
Applied Calculus
Section 4.3

Name ____________________

Date ____________________

Directions Each problem below is similar to the example with the same number in your textbook. After reading through an example in your textbook, or watching one of the videos of that example on MathTV.com, try the matched problem to check your progress in this section. You can use the QR code at the right to view the videos for this section on your smart phone, or tablet. If your smart device does not have a camera, but can access the internet, go to m.mathtv.com to view the videos for this worksheet.

1. For the function $f(x) = 5x^4 \ln x$, find
   a. $f'(x)$
   b. an equation of the line tangent at $x = e$.

2. Find $f'(x)$ for $f(x) = \ln(6x^3 - 7x + 2)$.

3. Find $f'(x)$ for $f(x) = \ln(5x + 6)^4$.

4. Find $f'(x)$ for $f(x) = [\ln(4x^2)]^5$.

5. Find $f'(x)$ for $f(x) = \dfrac{8}{\ln(6x^3)}$.

SECTION
4.3

6. Find y' for $x^2y + y^2 = 3$.

Matched Problems

Applied Calculus
Section 4.4

Name ______________________

Date ______________________

Directions Each problem below is similar to the example with the same number in your textbook. After reading through an example in your textbook, or watching one of the videos of that example on MathTV.com, try the matched problem to check your progress in this section. You can use the QR code at the right to view the videos for this section on your smart phone, or tablet. If your smart device does not have a camera, but can access the internet, go to m.mathtv.com to view the videos for this worksheet.

1. Find $f'(x)$ for $f(x) = 7x^4e^x$.

2. a. Find the equation of the line tangent to the curve $f(x) = 5xe^x$ at the point (0, 0).

b. Find the equation of the line tangent to the curve $f(x) = axe^x$ at the point (0, 0).

3. Find $f'(x)$ for

a. $f(x)` = e^{4x^3-7x}$

b. $f(x) = e^{(4x^3-7x)^2}$

4. Find $f'(2)$ for $f(x) = \ln(5x + e^{-0.3x})$.

5. Find $f''(x)$ for $f(x) = e^{3x^4-8}$.

6. Find $f'(x)$ for $f(x) = 5e^{-2x}\ln 6x$.

7. Find $f'(0)$ for $f(x) = \dfrac{e^x - 3}{e^x + 3}$.

8. Find each derivative.

a. $\dfrac{d}{dx}[e^{9x}]$

b. $\dfrac{d}{dx}[e^{-3x}]$

c. $\dfrac{d}{dx}[e^{-0.01x}]$

d. $\dfrac{d}{dx}[10e^{-0.5x}]$

9. Repeat example 9 using the function

$$R(t) = 375{,}000e^{-0.29t}$$

SECTION 4.4

10. Repeat example 10 using the function

$$A(t) = 5e^{-0.0158t}$$

Matched Problems
Applied Calculus
Section 5.1

Name ____________________

Date ____________________

Directions Each problem below is similar to the example with the same number in your textbook. After reading through an example in your textbook, or watching one of the videos of that example on MathTV.com, try the matched problem to check your progress in this section. You can use the QR code at the right to view the videos for this section on your smart phone, or tablet. If your smart device does not have a camera, but can access the internet, go to m.mathtv.com to view the videos for this worksheet.

1. Integrate the functions

a. $\int 7dx$ **b.** $\int dr$

2. Evaluate each indefinite integral

a. $\int x^6\, dx$ **b.** $\int x^{5/4}\, dx$ **c.** $\int \frac{1}{x^3}\, dx$

3. Evaluate each indefinite integral

a. $\int 10x^4\, dx$ **b.** $\int 7x^3\, dx$

4. Evaluate each indefinite integral

a. $\int (5x^3 - 8x + 4)\, dx$ **b.** $\int (8x^{1/5} - 3)\, dx$

5. Evaluate each indefinite $\int (4e^x + 3x^2)\,dx$

6. Evaluate each indefinite integral

a. $\int \frac{5}{x}\,dx$

b. $\int \left(\frac{6}{x} - 2e^x - 5\right) dx, x > 0$

c. $\int \frac{3x^2 - 8x + 4}{x}\,dx, x > 0$

SECTION 5.1

7. Find $f(x)$ if $f'(x) = 4x^3 + 9x^2 + 1$ and $f(2) = 42$.

8. The marginal cost function for a manufacturer is given by $C'(x) = -0.02x + 51$. If at the production level of 200 units, the cost is known to be $13,840, find

a. the cost function, $C(x)$, and

b. the cost of producing 300 units of the product.

Matched Problems

Applied Calculus
Section 5.2

Name ____________________

Date ____________________

Directions Each problem below is similar to the example with the same number in your textbook. After reading through an example in your textbook, or watching one of the videos of that example on MathTV.com, try the matched problem to check your progress in this section. You can use the QR code at the right to view the videos for this section on your smart phone, or tablet. If your smart device does not have a camera, but can access the internet, go to m.mathtv.com to view the videos for this worksheet.

1. Evaluate $\int \frac{18x + 8}{9x^2 + 8x - 3}\, dx$

2. Evaluate each indefinite integral.

a. $\int \frac{3x + 1}{3x^2 + 2x}\, dx$ **b.** $\int \frac{24x + 8}{3x^2 + 2x + 5}\, dx$

3. Evaluate each indefinite integral.

a. $\int 4e^{5x-8}\, dx$ **b.** $\int (4x - e^{-2x})\, dx$

4. Evaluate each indefinite integral.

a. $\int 12xe^{6x^2+1}\, dx$ **b.** $\int (16x + 10)e^{4x^2+5x}\, dx$

5. Evaluate the indefinite integral.

$$\int (10x - 7)(5x^2 - 7x)^4\, dx$$

6. Evaluate the indefinite integral.

$$\int \frac{8x + 2}{\sqrt[3]{2x^2 + x}}\, dx$$

SECTION
5.2

7. Evaluate the indefinite integral.

$$\int x(x - 5)^{1/3}\, dx$$

8. Evaluate each indefinite integral.

a. $\int e^{4x}\, dx$

b. $\int e^{\frac{3}{5}x}\, dx$

Matched Problems
Applied Calculus
Section 5.3

Name ______________________

Date ______________________

Directions Each problem below is similar to the example with the same number in your textbook. After reading through an example in your textbook, or watching one of the videos of that example on MathTV.com, try the matched problem to check your progress in this section. You can use the QR code at the right to view the videos for this section on your smart phone, or tablet. If your smart device does not have a camera, but can access the internet, go to m.mathtv.com to view the videos for this worksheet.

1. **a.** $\int_0^3 (x + 4)\, dx$

b. $\int_1^5 (3x^2 - 2x + 7)\, dx$

2. **a.** $\int_4^8 \frac{5}{(x - 2)}\, dx$

b. $\int_2^4 \frac{3}{(5 - x)}\, dx$

3. Do problem 2a using the method used in example 3.

Matched Problems

Applied Calculus
Section 5.4

Name ______________________

Date ______________________

Directions Each problem below is similar to the example with the same number in your textbook. After reading through an example in your textbook, or watching one of the videos of that example on MathTV.com, try the matched problem to check your progress in this section. You can use the QR code at the right to view the videos for this section on your smart phone, or tablet. If your smart device does not have a camera, but can access the internet, go to m.mathtv.com to view the videos for this worksheet.

1. Use five rectangles to approximate the area under the curve

$$f(x) = 0.4x^3 - 2.1x^2 + 20$$

from $x = 1$ to $x = 7$.

2. Use a definite integral to find the area under the curve

$$f(x) = 0.4x^3 - 2.1x^2 + 20$$

from $x = 1$ to $x = 7$.

3. Find the value of $\int_{6}^{10} (x^2 - 13x + 36)\, dx$.

4. Use a definite integral to find the area under the curve

$$f(x) = x^2 - 13x + 36$$

from $x = 6$ to $x = 10$.

SECTION

5.5

Matched Problems

Applied Calculus
Section 5.5

Name ____________________

Date ____________________

Directions Each problem below is similar to the example with the same number in your textbook. After reading through an example in your textbook, or watching one of the videos of that example on MathTV.com, try the matched problem to check your progress in this section. You can use the QR code at the right to view the videos for this section on your smart phone, or tablet. If your smart device does not have a camera, but can access the internet, go to m.mathtv.com to view the videos for this worksheet.

1. Evaluate the improper integral.
$\int_1^{\infty} \frac{1}{x^{3/2}}\, dx$

2. Evaluate $\int_{-\infty}^{4} \frac{1}{(x-5)^2}\, dx$.

SECTION 5.5

3. Evaluate the improper integral.
$\int_{-\infty}^{\infty} x^{1/3}\, dx$

4. Determine whether or not the integral $\int_{-2}^{2} \frac{1}{x^3}\, dx$ converges.

5. In certain areas of the ocean, there are fractures in the earth's crust where lava escapes through fissure eruptions, quickly cools, and hardens on the ocean floor. Suppose a certain fracture is closing up such that the rate of escaping lava in kilograms per year is given by

$$f(t) = \frac{44{,}000}{(2.2 + t)^2}$$

If the lava eruption continues indefinitely, how many kilograms of lava will have escaped through the fracture?

6. Suppose the concentration of a particular drug that is administered intravenously is given by $C_i(t) = 80e^{-0.4t}$. The concentration of that drug when administered orally is given by $C_0(t) = 100(e^{-0.4t} - e^{-1.0t})$. Determine the bioavailability, *F*, of the drug.

Matched Problems
Applied Calculus
Section 5.6

Name ____________________

Date ____________________

Directions Each problem below is similar to the example with the same number in your textbook. After reading through an example in your textbook, or watching one of the videos of that example on MathTV.com, try the matched problem to check your progress in this section. You can use the QR code at the right to view the videos for this section on your smart phone, or tablet. If your smart device does not have a camera, but can access the internet, go to m.mathtv.com to view the videos for this worksheet.

1. Evaluate $\int x^3 \ln x \, dx$.

2. Evaluate $\int xe^{3x} \, dx$.

3. Evaluate $\int x^2 \ln x \, dx$.

4. A manufacturer estimates that t months from now, the number N (in thousands) of units of one of its products that can be produced each month is approximated by

$$N(t) = 210te^{-0.7t}$$

Find the function that estimates the total number of units that can be produced if the total production initially zero.

SECTION
5.6

Matched Problems
Applied Calculus
Section 6.1

Name ____________________

Date ____________________

Directions Each problem below is similar to the example with the same number in your textbook. After reading through an example in your textbook, or watching one of the videos of that example on MathTV.com, try the matched problem to check your progress in this section. You can use the QR code at the right to view the videos for this section on your smart phone, or tablet. If your smart device does not have a camera, but can access the internet, go to m.mathtv.com to view the videos for this worksheet.

1. Find the area bounded by the line $f(x) = -3x + 26$, the curve $g(x) = x^2 - 10x + 32$, and the two vertical lines $x = 1$ and $x = 6$.

2. Find the area of the region bounded by the two curves $f(x) = 2x^2 - 28x + 99$ and $g(x) = -2x^2 + 12x + 15$, and the two vertical lines $x = 4$ and $x = 6$.

3. Find the area of the region that is completely bounded by the two curves

$f(x) = -x^2 + 8x + 70$ and $g(x) = x^2 + 10x + 30$

4. Find the area of the region that is completely bounded by the two curves

$$f(x) = x^3 - 5x + 20 \text{ and } g(x) = 4x + 20$$

5. Over the next 10 years, one company's financial analyst predicts profits will be given by $f(x) = 80e^{0.10x}$ millions of dollars and another company financial analyst predicts profits will be given by $g(x) = 80e^{0.08x}$ millions of dollars. What is the total difference in the estimations of profit made by these analysts? Round your answer to the nearest million dollars

Matched Problems
Applied Calculus
Section 6.2

Name ______________________

Date ______________________

Directions Each problem below is similar to the example with the same number in your textbook. After reading through an example in your textbook, or watching one of the videos of that example on MathTV.com, try the matched problem to check your progress in this section. You can use the QR code at the right to view the videos for this section on your smart phone, or tablet. If your smart device does not have a camera, but can access the internet, go to m.mathtv.com to view the videos for this worksheet.

1. The demand for a particular item is modeled by the function $D(x) = 1200 - 4x^2$. Find and interpret the consumer's surplus if the market price (equilibrium price) is \$300.

2. The demand and supply for a particular item is modeled by the functions $D(x) = 450 - 0.09x^2$ and $S(x) = 0.05x^2$, respectively. Find and interpret both the consumer's and producer's surplus.

3. The monthly demand x for a particular item is related to the price of the item by the function $D(x) = -0.002x^2 + 174$ dollars and the supply of the item is modeled by the function $S(x) = 0.0006x^2 + 0.44x + 50$ dollars. Find and interpret both the consumer's and producer's surplus.

Matched Problems

Applied Calculus

Section 6.3

Name ____________________

Date ____________________

Directions Each problem below is similar to the example with the same number in your textbook. After reading through an example in your textbook, or watching one of the videos of that example on MathTV.com, try the matched problem to check your progress in this section. You can use the QR code at the right to view the videos for this section on your smart phone, or tablet. If your smart device does not have a camera, but can access the internet, go to m.mathtv.com to view the videos for this worksheet.

1. Revenue is projected to flow continuously into a chain retail operation at a constant rate of $f(t) = 6{,}350{,}000$ dollars per year. As the money is received, it will be placed into a non-interest bearing account. Determine how much money will accumulate in the account over a 4 year period

2. If a one-time, lump-sum investment of $P = \$320{,}000$ is made into an account that pays 4% interest compounded continuously, find the amount of money that will have accumulated at the end of 5 years.

3. Suppose that at the end of each month, for 10 years, $1,000 is put into an annuity account paying 5% annual interest compounded continuously. Approximate the future value of the annuity.

4. Approximate the future value of the annuity in problem 3 over

a. a 1-year period

b. a 20-year period

5. How much money must be invested now at 3% interest compounded continuously so that \$30,000 will be available in 10 years?

6. XYZ Publishing can buy a machine that is expected to increase its net income by \$250,000 each year for the 7-year life of the machine. XYZ also estimates that for the next 7 years, money will be worth 4% compounded continuously. Find and interpret
   a. The total money flow over the 7-years life of the machine.
   b. The future value of the money stream.
   c. The present value of the money stream.

SECTION 6.3

7. Rental income from a piece of property, upon which there is an indefinite lease, flows in at the rate of \$144,000 per year. The income is invested immediately into an account paying 5.2% annual interest compounded continuously. Find and interpret the present value of the income flow.

Matched Problems

Applied Calculus
Section 6.4

Name ______________________

Date ______________________

Directions Each problem below is similar to the example with the same number in your textbook. After reading through an example in your textbook, or watching one of the videos of that example on MathTV.com, try the matched problem to check your progress in this section. You can use the QR code at the right to view the videos for this section on your smart phone, or tablet. If your smart device does not have a camera, but can access the internet, go to m.mathtv.com to view the videos for this worksheet.

1. Show that the differential equation $\frac{dy}{dx} = 3x^3(y^3 + 5)$ is separable.

2. Solve the differential equation $(x^2 - 5)\frac{dy}{dx} = 5x$.

3. Solve the differential equation $y' = 4x^3y^2$ subject to the conditions that $y = -1$ when $x = 2$.

4. Solve the differential equation $y' = \frac{4x^2}{y^2}$ subject to the initial condition that $y = 2$ when $x = 1$.

5. Solve each differential equation.
   a. $y' = 3x^4y$ subject to the conditions that $y = e$ when $x = 0$.
   b. $y' = -2y$ subject to the conditions that $y = 3$ when $x = \ln 2$.

6. The growth of a new publishing company depends on the number subscriptions it gets according to the differential equation $N'(t) = 0.8\sqrt{N(t)}$, where $N(t)$ represents the number of subscribers at any time t and $N'(t)$ represents the rate at which the number of subscribers is changing at time t. Solve this differential equation if at the start of an advertising campaign, there are 100 subscribers and determine how many subscribers the company expects to have 12 months after the start of the advertising campaign.

Matched Problems
Applied Calculus
Section 6.5

Name ____________________

Date ____________________

Directions Each problem below is similar to the example with the same number in your textbook. After reading through an example in your textbook, or watching one of the videos of that example on MathTV.com, try the matched problem to check your progress in this section. You can use the QR code at the right to view the videos for this section on your smart phone, or tablet. If your smart device does not have a camera, but can access the internet, go to m.mathtv.com to view the videos for this worksheet.

1. The value of real estate in a city is decreasing, without limitations, at a rate proportional to its current value. A piece of property in the city on January 1, 2009 appraised at $840,000 and, three years later on January 1, 2012 at $610,000. According to this model, at what value will this piece of property be appraised in January 2015?

Source: Los Angeles County Office of the Assessor

2. A bar of carbon steel is removed from a heat source when its temperature is 723° C. It is put into a cooler environment in which the surrounding air temperature is 90° C. After 10 minutes, the steel bar has cooled to 690° C. Write the function that gives the temperature of the steel bar at any time *t* and then find the temperature of the bar after 5 hours (300 minutes).

Source: Gowelding

3. Information can spread quickly throughout a social network. Suppose that a particular video sharing social network is comprised of 370,000 people. On a particular day, 120 people learn about a piece of information that would be of interest to the entire group and they release the information to all group members. Five hours later, 8,500 people in the network have heard the information. Furthermore, suppose that the number of people who have heard about the information increases at a rate proportional to the difference between the total number of people in the network and the number of people who have already heard about the information.
   a. Write a function that gives the total number of people in the network who have heard the information at any time t.
   b. How many people in the network will have heard about the information 24 hours after the time it was released.
   c. Set $P(t) = 369{,}880$ and solve for t to find when all the people in the network will have heard the information. Use 369,880 and not 370,000 because 120 people new about the information before the information was released. $370{,}000 - 120 = 369{,}880$

   Source: Science Daily

4. Thirty-six months ago, 410 plants in a forested area had a particular disease. In a study reported today, 1,550 plants had the disease. The number of plants contracting the disease appears to be increasing at a rate that is proportional to the current size of the number of plants with the disease in the area and the difference between the total number of plants that are susceptible to this disease and the current number of plants with the disease. There are 14,000 plants in the area.
   a. Write a function that gives the number of plants with the disease at any time t.
   b. If this model is correct, how many would be expected to have this disease 5 years from now. (Five years from now will make $t = 96$ since $t = 0$ is 36 months ago and 5 years is 60 months.)

   Source: American Phytopathological Society

SECTION 6.5

Matched Problems
Applied Calculus
Section 6.6

Name ____________________

Date ____________________

Directions Each problem below is similar to the example with the same number in your textbook. After reading through an example in your textbook, or watching one of the videos of that example on MathTV.com, try the matched problem to check your progress in this section. You can use the QR code at the right to view the videos for this section on your smart phone, or tablet. If your smart device does not have a camera, but can access the internet, go to m.mathtv.com to view the videos for this worksheet.

1. Show that $f(x) = \frac{4}{81}x^3$ is a valid probability density function for each x in $[0, 3]$ given that $f(x) = 0$ for every x outside of $[0, 3]$.

2. Find the probability that a randomly selected value of x lies in the interval $[0, 2]$ if $f(x) = \dfrac{4x^3}{81}$. Round the probability to 4 decimal places.

3. A bus arrives at a bus stop every 15 minutes. Assume that without failure, busses really do arrive every 15 minutes. Suppose you arrive at the bus stop and see that the bus is not there. What is the probability that you have to wait less than 5 minutes for the next bus to arrive?

SECTION 6.6

4. The research and development department of an audio equipment manufacturer has determined that its new wave-shaper circuit has a lifespan (in months) that is exponentially distributed with probability density function

$$f(x) = 0.01e^{-0.01x}$$

Find the probability that a new and randomly selected wave-shaper made by this manufacturer will last at most 75 months under continuous use.

Source: Sonic Lion

5. A machine is adjusted to dispense 6 ounces of coffee into a 7-ounce cup. The amount of coffee dispensed is normally distributed with mean $\mu = 6$ ounces and standard deviation $\sigma = 0.5$ ounces. Find the probability that a randomly selected cup is filled with between 5.0 and 5.7 ounces of coffee into a 7-ounce cup?

Matched Problems
Applied Calculus
Section 7.1

Name ____________________

Date ____________________

Directions Each problem below is similar to the example with the same number in your textbook. After reading through an example in your textbook, or watching one of the videos of that example on MathTV.com, try the matched problem to check your progress in this section. You can use the QR code at the right to view the videos for this section on your smart phone, or tablet. If your smart device does not have a camera, but can access the internet, go to m.mathtv.com to view the videos for this worksheet.

1. The Ventus Potestatem wind energy company manufactures and sells 5 kW, 35 kW, and 50 kW wind turbines to small farm and residences. The 5 kW turbines sell for $22,500 each, the 35 kW turbines for $140,000 each, and the 50 kW turbines for $190,000 each. The company's revenue for the sale of these turbines is modeled by the function of three variables

$$R(x, y, z) = 22{,}500x + 140{,}000y + 190{,}000z$$

where x represents the number of 5 kW turbines sold, y represents the number of 35 kW turbines sold, and z represents the number of 50 kW turbines sold.

What is the company's revenue if it sells thirty of the 5 kW turbines, forty of the 35 kW turbines, and 20 of the 50 kW turbines?

Source: Windustry

SECTION 7.1

2. A company models is production using the Cobb-Douglas production function

$$f(x, y) = 650x^{4/5}y^{1/5}$$

a. Determine the company's level of production when 243 units of labor and 1,024 units of capital are used.

b. Estimate the change in the level of production if the number of units of labor is decreased by 1, from 243 to 242, and the number of units of capital is increased by 10, from 1,024 to 1,034.

Matched Problems
Applied Calculus
Section 7.2

Name ____________________

Date ____________________

Directions Each problem below is similar to the example with the same number in your textbook. After reading through an example in your textbook, or watching one of the videos of that example on MathTV.com, try the matched problem to check your progress in this section. You can use the QR code at the right to view the videos for this section on your smart phone, or tablet. If your smart device does not have a camera, but can access the internet, go to m.mathtv.com to view the videos for this worksheet.

1. For the partial derivative

$$\frac{\partial}{\partial x} f(x, y) = 16x - 20y + 7$$

compute and interpret the meaning of

a. $\frac{\partial}{\partial x} f(5, 2)$

b. $\frac{\partial}{\partial x} f(1, 2)$

2. For the function

$$f(x, y) = 5x^4 - 2y^3 + 4x + y$$

a. Find $\frac{\partial f}{\partial x}$

b. Find $\frac{\partial f}{\partial y}$

c. Evaluate and interpret $\frac{\partial f}{\partial x} f(2, 3)$

d. Evaluate and interpret $\frac{\partial f}{\partial y} f(2, 3)$

SECTION 7.2

3. For the function

$$f(x, y) = x^3 + 3x^3y^2 - 6y^2$$

a. Find $f_x(x, y)$

b. Find $f_y(x, y)$

c. Evaluate and interpret $f_x(4, 1)$

d. Evaluate and interpret $f_y(4, 1)$

4. For the function

$$f(x, y) = (2x^5 - 7y^3)^6$$

a. Find f_x

b. Find f_y

5. For the function

$$f(x, y, z) = e^{2y+5z} + 8\ln(xz)$$

a. Find f_x
b. Find f_y
c. Find f_z

6. Find all the points (x, y) for which both f_x and f_y equal zero, where $f(x, y) = 3x^2 + 2y^2 - 2xy - 5$

7. Find the second partial derivatives of

$$f(x, y) = 2x^4 - 3x^3y - 4xy^3 + 3y^4$$

8. The Gigas Lepus carpet company uses two chemicals to dry clean their carpets. They use x units of chemical A and y units of chemical B. The amount of residue left in the cleaning apparatus is given by the function

$$R(x, y) = -0.03x^2 + 0.05xy + 0.01y^2$$

Currently, the company uses 20 units of chemical A and 8 units of chemical B in their cleaning solution.

a. Find the rate at which the residue is changing as the number of units of chemical B changes.
b. Determine if this rate is changing at an increasing or decreasing rate.

Matched Problems
Applied Calculus
Section 7.3

Name ____________________

Date ____________________

Directions Each problem below is similar to the example with the same number in your textbook. After reading through an example in your textbook, or watching one of the videos of that example on MathTV.com, try the matched problem to check your progress in this section. You can use the QR code at the right to view the videos for this section on your smart phone, or tablet. If your smart device does not have a camera, but can access the internet, go to m.mathtv.com to view the videos for this worksheet.

1. Find and interpret the critical points of the function

$$f(x, y) = 2x^2 + \frac{3}{2}y^2 - 12x - 3y + 10$$

2. Find and interpret the critical points of the function

$$f(x, y) = \frac{3}{2}x^2 + 2xy + \frac{1}{2}y^2 - 7x - 32y + 40$$

3. Find and interpret the critical points of the function

$$f(x, y) = \frac{5}{2}x^2 + \frac{3}{4}y^4 - 6y^2 - 5x$$

4. Find and interpret the critical points of the function

$$f(x, y) = 6x + 5y + 6$$

5. Locate the relative extrema, if any exist, of

$$f(x, y) = 3x^2 + y^2 - 18x + 20y + 8$$

6. Locate the relative extrema, if any exist, of

$$f(x, y) = x^3 + \frac{4}{3}y^3 - 2y^2 - 27x + 9$$

7. The Drut company manufacturers and sells stainless steel wire in two locations, Maryland and Georgia. In order to sell x units of wire in Maryland, Drut sets the price per unit at $120 - \frac{x}{25}$ dollars, and in order to sell y units of wire in Georgia, Drut sets the price per unit at $100 - \frac{y}{15}$ dollars. The cost of producing all $x + y$ units is $70{,}000 + 10(x + y)$. How many units of wire should be produced for Maryland and Georgia, respectively to maximize Drut's profit?

Matched Problems

Applied Calculus
Section 7.4

Name ______________________

Date ______________________

Directions Each problem below is similar to the example with the same number in your textbook. After reading through an example in your textbook, or watching one of the videos of that example on MathTV.com, try the matched problem to check your progress in this section. You can use the QR code at the right to view the videos for this section on your smart phone, or tablet. If your smart device does not have a camera, but can access the internet, go to m.mathtv.com to view the videos for this worksheet.

1. Use the substitution method to find and classify the relative extreme points of the function

$$f(x, y) = x^2 - 8y^2 + 4x + 14y + 20$$

subject to the constraint

$$x - 3y = -3$$

2. Use the method of Lagrange multipliers to find the minimum of the function

$$f(x, y) = 5x^2 + 5y^2$$

subject to the constraint

$$2x - y = 10$$

3. The production of a manufacturer is given by the Cobb-Douglas production function $f(x, y) = 30x^{2/5}y^{3/5}$, where x represents the number of units of labor (in hours) used and y the number of units of capital (in dollars) invested. Labor costs \$30 per hour and there are 8 hours in a working day and 250 working days in a year. The manufacturer has designated \$10,000,000 this year for labor and capital. How should the money be allocated to labor and capital so as to maximize the year's productivity?

Matched Problems
Applied Calculus
Section 7.5

Name ______________________

Date ______________________

Directions Each problem below is similar to the example with the same number in your textbook. After reading through an example in your textbook, or watching one of the videos of that example on MathTV.com, try the matched problem to check your progress in this section. You can use the QR code at the right to view the videos for this section on your smart phone, or tablet. If your smart device does not have a camera, but can access the internet, go to m.mathtv.com to view the videos for this worksheet.

1. The productivity of a company is modeled by the Cobb-Douglas production function

$$f(x, y) = 64x^{3/8}y^{5/8}$$

where x represents the number of units of labor used and y the number of units of capital used. Approximate the change in output if the number of units of labor is increased from 6,561 units to 6,566 units and capital is decreased from 256 units to 250 units.

Matched Problems

Applied Calculus
Section 7.6

Name ______________________

Date ______________________

Directions Each problem below is similar to the example with the same number in your textbook. After reading through an example in your textbook, or watching one of the videos of that example on MathTV.com, try the matched problem to check your progress in this section. You can use the QR code at the right to view the videos for this section on your smart phone, or tablet. If your smart device does not have a camera, but can access the internet, go to m.mathtv.com to view the videos for this worksheet.

1. If $f_y(x, y) = 12xy^3$, find $f(x, y)$ by evaluating $\int 12xy^3\, dy$. Use $h(x)$ to represent the "constant" of integration.

2. In Example 2, we found the volume of the space under the surface using

$$V = \int_a^b \int_{g_1(x)}^{g_2(x)} f(x, y)\, dydx$$

Find the volume of the same space under the same surface $ff(x, y) = x + 4y$ and over the same rectangular region bounded by $2 \le x \le 5$ and $3 \le y \le 10$ but this time using

$$V = \int_c^d \int_{h_1(y)}^{h_2(y)} f(x, y)\, dxdy$$

3. Find the volume of the space under the surface

$$f(x, y) = 2x^3 + 2xy^2$$

and over the region bounded by the curve $g_1(x) = x^2$, the line $g_2(x) = x$, and between $x = 0$ and $x = 1$. Use the order of integration *dydx*.

4. Show that the same volume from problem 3 can be found by changing the order of integration to *dxdy*.

5. Quartz crystals are thin slices of quartz that, when placed between two electrodes and an alternating current applied, can be made to vibrate at specified frequencies. They are commonly used in radio transmitters for tracking salmon and whales and other wildlife, as well as meters for gas pumps and instruments at the bottom of oil wells. A manufacturer of quartz crystals produces crystals that are to be 12.5 mm in diameter and 0.305 mm in thickness. Because of some variability in the production process, the crystals have diameters that are uniformly distributed 12.45 and 12.55 mm and thicknesses that are uniformly distributed between 0.300 and 0.310. The cuts for the diameter and thickness are independent of each other. What is the probability that a randomly quartz crystal produced by this company will have a diameter between 12.49 and 12.51 mm and a thickness between 0.303 and 0.307 mm?

Source: Structure Probe, Inc

6. How many molds are on a 10" high and 20" long surface if the probability density function for the mold is given by

$$P(x, y) = 300e^{-0.2x-0.3y}$$

where x and y are in inches and P is in millions.

Matched Problems
Applied Calculus
Section 7.7

Name ______________________

Date ______________________

Directions Each problem below is similar to the example with the same number in your textbook. After reading through an example in your textbook, or watching one of the videos of that example on MathTV.com, try the matched problem to check your progress in this section. You can use the QR code at the right to view the videos for this section on your smart phone, or tablet. If your smart device does not have a camera, but can access the internet, go to m.mathtv.com to view the videos for this worksheet.

1. A study indicates that the number of people who believe a known untrue "fact" is modeled by the function

$$N = 10{,}000te^{-0.02t^2}$$

where t is in days. What is the average number of people per day who believe the untrue "fact" over the next year?

2. Each month a manufacturer of farm tractor wheels sells between 0 and 100 wheels of type x and between 0 and 100 wheels of type y. The manufacturer estimates the monthly profit using the function

$$P(x, y) = 6{,}000 - 2x^2 + 1{,}000x + 4(y - 100)^2$$

Find the manufacturer's average monthly profit.

Source: farmwheel.com

Worksheet Problems — Name

Applied Calculus — Date

Section 7.7

[illegible]

[illegible]

2. Each month a manufacturer [illegible] sells between 0 and 100 [illegible] of type x and between 0 and 100 [illegible] of type y. The manufacturer estimates the monthly profit using the function

$$P(x, y) = [illegible]$$

Find the manufacturer's average monthly profit.

[illegible]

Matched Problems
Applied Calculus
Answers

Section 1.1

1.

x	Function Rule $y = 8x$	y
0	$y = 8(0)$	0
10	$y = 8(10)$	80
20	$y = 8(20)$	160
30	$y = 8(30)$	240
40	$y = 8(40)$	320

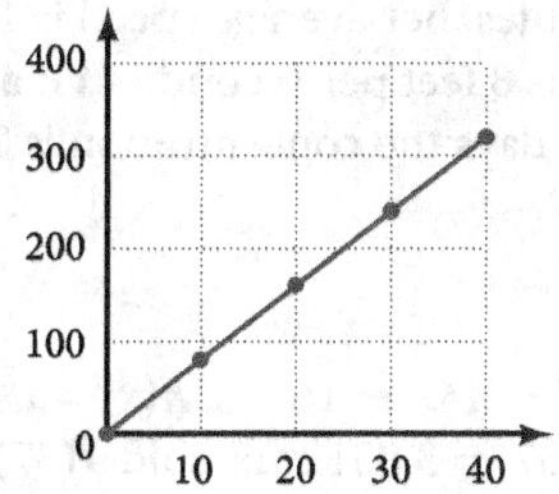

2. Domain: $\{0 \le x \le 40\}$, Range: $\{0 \le x \le 320\}$

3.

Time (sec) t	Function Rule $h = 48t - 16t^2$	Distance h
0	$h = 48(0) - 16(0)^2$	0
1	$h = 48(1) - 16(1)^2$	32
1.5	$h = 48(1.5) - 16(1.5)^2$	36
2	$h = 48(2) - 16(2)^2$	32
3	$h = 48(3) - 16(3)^2$	0

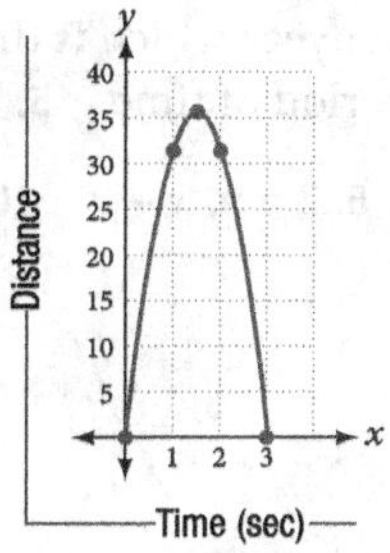

4. This relation is not a function.

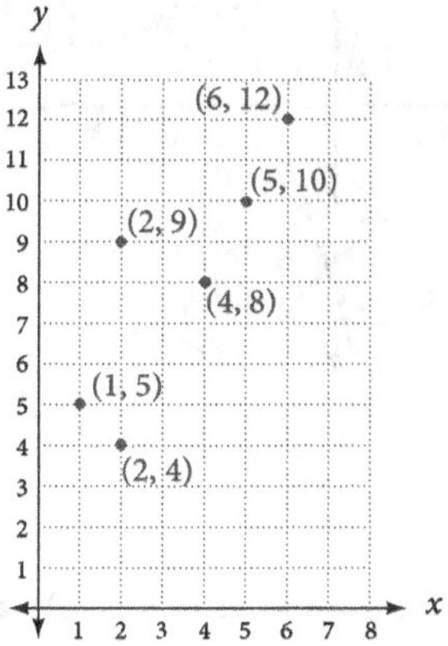

5. No, this is not the graph of a function.

y	Function Rule $x = y^2 - 4$	x
−3	$x = (-3)^2 - 4$	5
−2	$x = (-2)^2 - 4$	0
−1	$x = (-1)^2 - 4$	−3
0	$x = (0)^2 - 4$	−4
1	$x = (1)^2 - 4$	−3
2	$x = (2)^2 - 4$	0
3	$x = (3)^2 - 4$	5

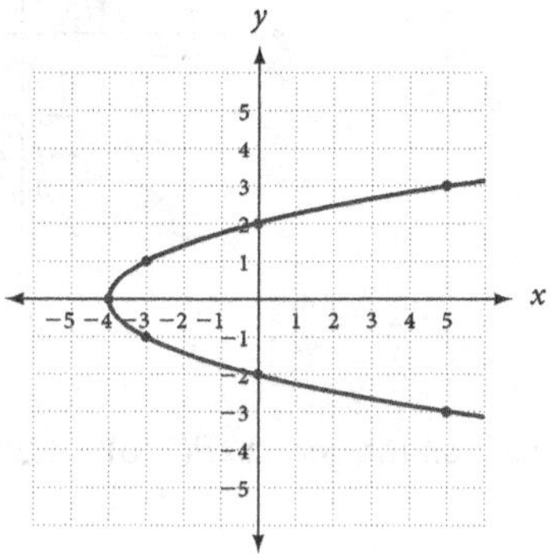

6. Yes, this is a graph of a function.

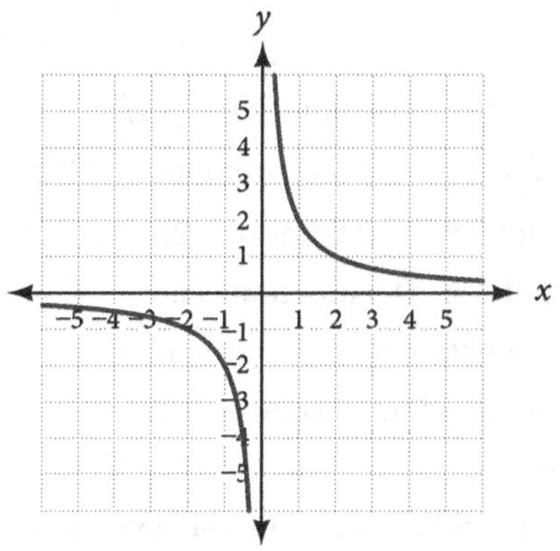

7. Yes, this is a graph of a function.

x	Function Rule $y = \|x\| - 3$	y
−3	$y = \|-3\| - 3$	0
−1	$y = \|-1\| - 3$	−2
0	$y = \|0\| - 3$	−3
1	$y = \|1\| - 3$	−2
3	$y = \|3\| - 3$	0

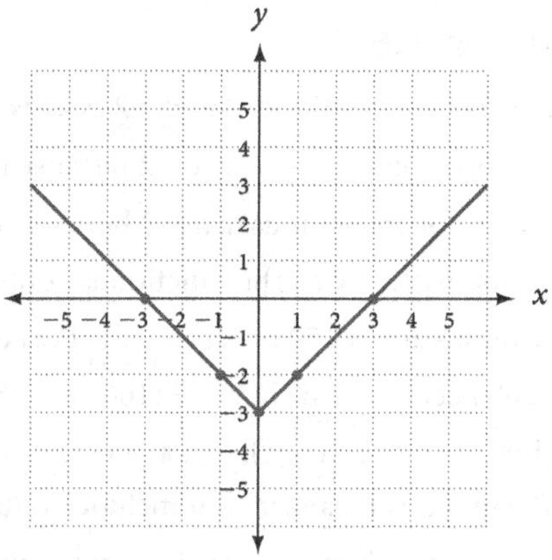

Matched Problems
Applied Calculus
Answers

8. a. 0 **b.** 40 **c.** 84 **9. a.** 11 **b.** 22 **c.** −3 **d.** 1 **e.** $2a + 1$ **f.** $a^2 - 3$
10. a. $s(8) = 11$; it means that if Lorena runs a mile in 8 minutes, her average speed is 11 feet per second **b.** $s(11) = 8$; it means that if Lorena runs a mile in 11 minutes, her average speed is 8 feet per second. **11. a.** $C(5) = 40$ ng/mL; after 5 days the concentration in 40 ng/mL. **b.** $C(10) = 20$ ng/mL; after 10 days the concentration is 20 ng/mL. **12.** $C(5) = 10\pi = 31.4$ inches, $A(5) = 25\pi = 78.5$ square inches

Section 1.2

1. a. $3x^2 - 7x - 6$ **b.** $g(x) = 3x^2 - 10x - 8$ **c.** $9x^3 - 24x^2 - 44x - 16$ **d.** $h(x) = x - 4$ **2. a.** 27 **b.** 14 **c.** −9 **d.** 9
3. $P = 6$ or $P = 7$ **4. a.** $x^2 + 3x - 4$ **b.** $x^2 - 5x + 4$ **5.** $s(37) = 3{,}851$ cans sold; $r(37) = 489$ cans recycled

Section 1.3

1. Slope: 3

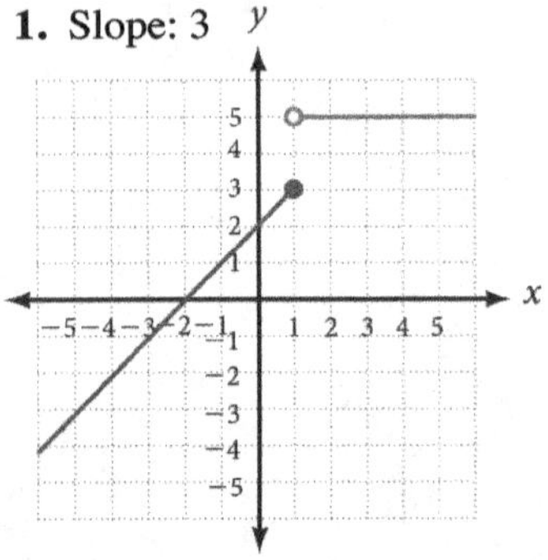

2. 3 **3.** 0 **4.** 8.4 cents per year which is the average rate of change of the price of a gallon of gasoline over a 20-year period of time. **5.** −4 feet per second **6.** −$3,187.50 per year
7. $40\pi \dfrac{\text{in}^2}{\text{in}} \approx 125.66 \dfrac{\text{in}^2}{\text{in}}$ **8.** 3 **9.** $x + a$ **10.** $y = \dfrac{2}{3}x + 1$ **11.** $m = \dfrac{4}{5}$, $b = -\dfrac{7}{5}$

12.

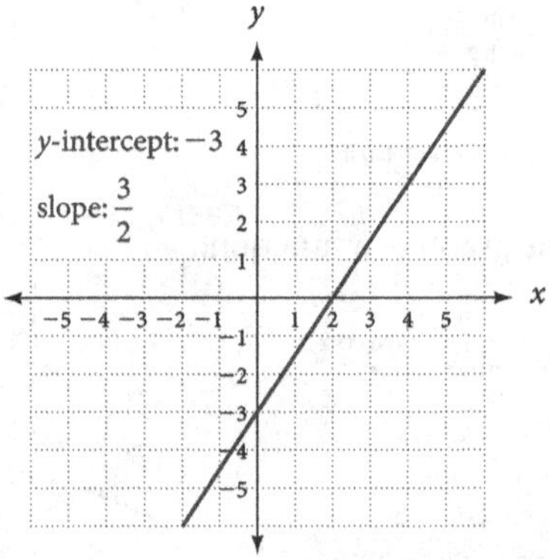

13. $y = 3x + 5$ **14.** $y = -2x + 9$

15.

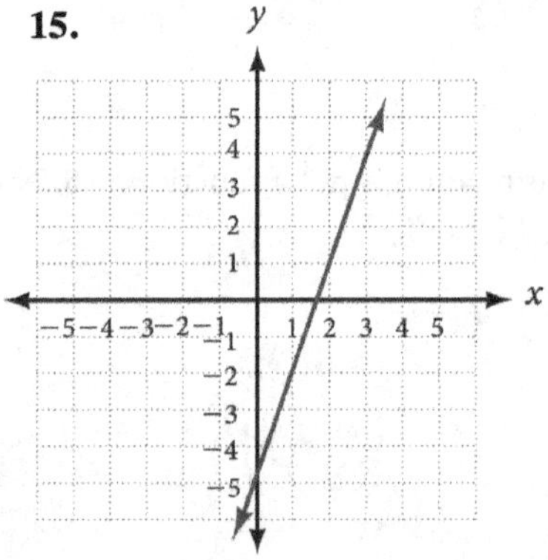

16. a. $0 \le x \le 350$, where x is an integer **b.** $P(140) = 1{,}200$ **c.** $P(150) = 1{,}300$ **d.** $P(160) = 1{,}250$

Section 1.4

1. 4 **2. a.** 1,200 **b.** 1,300 **c.** 1,250 **3.** −3 **4.** 9 **5.** 3 **6.** 8 **7.** 9 **8.** 2 **9.** 1 **10.** 8 **11. a.** 1 **b.** ∞ (or does not exist) **c.** 1
12. Does not exist **13.** −3 **14.** $\dfrac{1}{6}$ **15.** $\dfrac{5}{7}$ **16.** ∞ (or does not exist) **17.** 0 **18. a.** 0 **b.** 3 **c.** ∞ (or does not exist)
19. Vertical asymptote not in domain; horizontal asymptote: $W(x) = 160$

Section 1.5

1. a. $f(x)$ is not continuous at $x = 2$ because f(2) does not exist. **b.** $g(x)$ is not continuous at $x = 2$ because $\lim_{x \to 2} g(x) \neq g(2) = 3$ **c.** $h(x)$ is not continuous at $x = 2$ because $\lim_{x \to 2} h(x) \neq h(2) = 5$ **2. a.** $\lim_{x \to 2^-} f(x) = 3$ and $\lim_{x \to 2^+} f(x) = 3$ so $f(x)$ is continuous at $f(2)$ **b.** $g(x)$ is not continuous at $x = 2$ because $\lim_{x \to 2^-} g(x) = 5 \neq \lim_{x \to 2^+} g(x) = 3$ **c.** $h(x)$ is not continuous at $x = 2$ because $h(2)$ does not exist. **3. a.** The function is continuous at $x = 140$ because $\lim_{x \to 140} P(x) = P(140) = 1{,}200$. **b.** The function is not continuous at $x = 150$ because $\lim_{x \to 150} P(x)$ does not exist, or $\lim_{x \to 150^-} P(x) \neq \lim_{x \to 150^+} P(x)$. **c.** The function is continuous at $x = 160$ because $\lim_{x \to 160} P(x) = P(160) = 1{,}250$ **4.** No, the function is not continuous at $x = a$ because $\lim_{x \to a} f(x)$ does not exist, or $\lim_{x \to a^-} f(x) \neq \lim_{x \to a^+} f(x)$. **5. a.** $f(x)$ is continuous for all x $(-\infty, \infty)$ **b.** $g(x)$ is continuous for $(-\infty, -4) \cup (-4, 5) \cup (5, \infty)$
6. The graph is idealized as it includes values that are not whole numbers; the number of tickets sold can only be whole numbers. The actual graph would consist of unconnected, yet collinear points.

Matched Problems
Applied Calculus
Answers

Section 1.6

1. $\frac{1}{2} = 0.5$ **2.** 6 **3.** 7 **4.** $-2x - h + 13$ **5. a.** 16 **b.** 30 **c.** -54 **6. a.** 154 **b.** 130

5. At $t = \frac{1}{4}$ seconds, the instantaneous rate of change is 56 ft/s. At $t = 1$ second, the instantaneous rate of change is 32 ft/s.

Chapter 2
Section 2.1

1. a. The instantaneous rate of change is 6, which is a constant everywhere. **b.** This means the original function is linear. **2. a.** $f'(x) = 2x - 8$ gives the instantaneous rate of change of f with respect to x. **b.** $f'(3) = -2$ The slope of the tangent line when $x = 3$ is -2. **3. a.** For any value of x, if x increases by 1 unit in value from 3 to 4, then $f(x)$ will change by exactly 0 units in value. That is, a 1-unit increase in the x value produces no change at all in the $f(x)$ value. **b.** For any value of x, if x increases by 1 unit in value to $x + 1$, then $f(x)$ will change by exactly 0 units in value. That is, a 1-unit increase in the x value produces no change at all in the $f(x)$ value. **4. a.** $f'(x) = 7x^6$ **b.** $f'(x) = 10x^9$ **c.** $f'(x) = 1$

5. a. $f(16) = (\sqrt[4]{16})^3 = 2^3 = 8$ **b.** $f'(x) = \frac{3}{4x^{1/4}}$, and $f'(16) = \frac{3}{8}$ **6. a.** $f'(x) = \frac{-2}{x^3}$ **b.** $y = -2x + 3$ **7.** $f'(x) = \frac{-42}{x^8}$

8. $\frac{dy}{dx} = 7$ **9. a.** $y' = 9$ **b.** $f'(x) = -12$ **10.** $f'(x) = 12x^2 - 14x - 8$ **11. a.** $f(x) = 0$ when $x = -4, 5$ **b.** $f'(x) = 0$ when $x = \frac{1}{2}$

12. $\frac{8}{27}$ **13.** $\frac{dy}{dx} = \frac{3}{x^{3/4}}$. Then $\left.\frac{dy}{dx}\right|_{x=16} = \frac{3}{16^{3/4}} = \frac{3}{8}$ If the value of x increases by 1 unit, from 16 to 17, the value of y will increase by approximately $\frac{3}{8}$ units. **14. a.** 19,125 people **b.** 2,325 more people the next day.

Section 2.2

1. $f'(x) = 8x^3 + 10x$ **2. a.** $g(t) = 0$ when $t = -3, -1, 1, 3$ **b.** $g'(t) = 0$ when $t = 0, \pm\sqrt{5}$

3. a. $f'(x) = \frac{x^4 - 14x^2 - 8x - 15}{(x^2 - 3)^2}$ **b.** $f'(0) = \frac{-15}{9} = \frac{-5}{3}$

4. a. This quotient indicates that 5 years from now, the revenue predicted by analyst C will be 7 times as big as that predicted by analyst D. **b.** This quotient indicates that 5 years from now, the revenue predicted by analyst C will be increasing 7 times as fast as that predicted by analyst D. If the number of years from now increases by 1, from 5 to 6, the revenue predicted by analyst C will be increasing 7 times as fast as that predicted by analyst D. In 5 years, analyst C predicts the revenue will be 42 units and increasing at the rate of 14 units per year. Thus, in year 6, analyst C's predicted revenue will be $42 + 14 = 56$ units. Also in 5 years, analyst D predicts the revenue will be 6 units and increasing at the rate of 2 units per year. Thus, in year 6, analyst D's predicted revenue will be $6 + 2 = 8$ units. Note that in year 6, analyst C's predicted value will be $\frac{56}{8} = 7$ times that of analyst D's predicted value. **c.** This means that in the year 6, C's predicted value will grow to 3 times that of what D's predicted value is here in year 5. Currently, here in year 5, C's value is already 7 times that of D's, so in year 6, C's predicted value will be $7 \cdot 3 = 21$ times as big as D's predicted value.

$P'(x) = \frac{-20(5t^2 + 66t - 30}{(t^2 + 6)^2}$ **d.** From month 5 to month 6, the publisher can expect the profit to decrease by about $8,800. **e.** From month 10 to month 11, the publisher can expect the profit to decrease by about $2,000. **f.** For all values of $t > 1$, $P'(x) < 0$, demonstrating that the profit function, $P(t)$, is always decreasing. **5.** $y = -14x + 5$

Matched Problems
Applied Calculus
Answers

Section 2.3

1. a. $f'(x) = 10x^4 - 12x^3 + 14x + 5$ **b.** $f'(x) = 24x^2 - 2$ **2. a.** $f(2) = \frac{-80}{3}$ **3. a.** $g(t) = 0$ when $t = 0, -2, 2$

$f''(x) = 40x^3 - 36x^2 + 14$ $f''(x) = 48x$ **b.** $f'(2) = -32$ **b.** $g'(t) = 0$ when $t = \pm\frac{2\sqrt{3}}{3}$

$f'''(x) = 120x^2 - 72x$ $f'''(x) = 48$ **c.** $f''(2) = 16$ **c.** $g''(t) = 0$ when $t = 0$

$f^{(4)}(x) = 240x - 72$ $f^{(4)}(x) = 0$

4. $y' = \frac{-4}{x^3}$, $y'' = \frac{12}{x^4}$, $y''' = \frac{-48}{x^5}$ **5.** In the year 2008, donations to colitis researchers were \$42 million dollars. The first derivative is positive, telling us that in the year 2008, donations were increasing. But the second derivative is negative, telling us that the increase is slowing. So, in the year 2008, this information indicates that donations are increasing at a decreasing rate.
6. In the year 2013, the city can expect its health care costs for uninsured citizens to be decreasing but at an increasing rate.
7. Thirteen months after January 1, 2012, the company can expect its monthly energy costs to be decreasing at decreasing rate.

Section 2.4

1. a. $\frac{dy}{dx} = 15(5x - 1)^2$ **b.** When $x = 2$, $\frac{dy}{dx} = 1215 = \frac{1215}{1}$. If x increases by 1 unit, from 2 to 3, y will increase by approximately 1,215 units. **2. a.** $C'(p) = \frac{5}{3}p^{-\frac{1}{3}}$ **b.** The cost of production increases by about \$0.56.
3. a. $f'(x) = 8(5x + 3)(5x^2 + 6x - 8)^3$ **b.** $f'(x) = \frac{105}{(4 - 7x)^4}$ **c.** $f'(x) = \frac{5}{6(5x - 2)^{5/6}}$ **4.** $f'(-1) = \frac{5589}{4}$.
5. $f'(x) = \frac{x(7x + 30)}{3(x + 5)^{2/3}}$ **6.** $f'(x) = \frac{156(3x - 4)^5}{(5x + 2)^7}$

Section 2.5

1. a. $y' = \frac{-12x^2}{12y^3 + 1}$ **b.** At (3, 1), $y' = \frac{-108}{13}$ **2.** $y' = \frac{8x - 4x^3y^4}{4x^4y^3 + 5}$ **3.** $\frac{-160}{9}$; If t increases by 1 unit, when $\frac{dx}{dt} = 5$, $x = 2$, and $y = 3$, then the value of y can be expected to decrease by approximately 17.8 units. **4.** 10,205 sq. ft/hr; In the next hour the area of the oil spill can be expected to increase by approximately 10,205 sq. ft. **5** −22; At the time when the price of a precision instrument is \$400 and is increasing at the rate of \$1 per month, the company can expect to sell about 22 fewer instruments in the next month.

Chapter 3
Section 3.1

1. It appears that $f(x)$ increases on the intervals $(-\infty, -1.5)$ and $(2, \infty)$ that it decreases on the interval $(-1.5, 2)$.
2. $f(x) = 3x^2 - 15x$ is decreasing on $\left(-\infty, \frac{5}{2}\right)$ and increasing on $\left(\frac{5}{2}, \infty\right)$.

3. $g'(x) = 0$ so that $g(x)$ is constant. The function values are always 7. The function values neither increase nor decrease. The graph reflects this constant behavior.

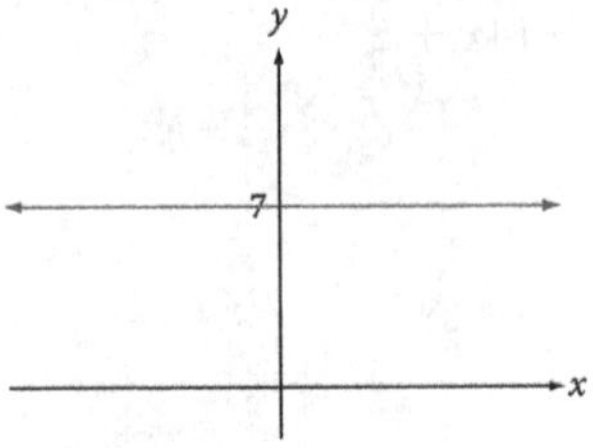

4.

Interval/Value	$f'(x)$	$f(x)$	Behavior of $f(x)$
$(-\infty, -6)$	(−)		$f(x)$ is decreasing
−6	0	−648	(−6, −648) is a relative minimum
(−6, 0)	(+)		$f(x)$ is increasing
0	0	0	(0, 0) is a relative maximum
(0, 6)	(−)		$f(x)$ is decreasing
6	0	−648	(6, −648) is a relative minimum
$(6, \infty)$	(+)		$f(x)$ is increasing

Matched Problems
Applied Calculus
Answers

5.

Interval/Value	$f'(x)$	$f(x)$	Behavior of $f(x)$
−2		1	
(−2, 1)	(+)		$f(x)$ is increasing
−1	Undefined	3	(−1, 3) is a absolute maximum
(−1, 0)	(−)		$f(x)$ is decreasing
0	0	1	(0, 1) is a relative minimum
(0, 1)	(+)		$f(x)$ is increasing
1	Undefined	3	(1, 3) is an absolute maximum
(1, 2)	(−)		$f(x)$ is decreasing
2		0	(2, 1) is an absolute minimum

6. Accept all 3 claims.
7. Absolute max at (8, 4.44) and absolute min at (0, 1.29). (Note: Should round to same number of digits.)
8. No critical values; no critical extrema.

Section 3.2

1. From 4:00 p.m. to 8:00 p.m., efficiency is increasing at an increasing rate. From 8:00 p.m. to midnight, efficiency is increasing but at a decreasing rate. From midnight to 4:00 a.m., efficiency is decreasing at an increasing rate. The maximum efficiency occurs at midnight, 8 hours after the start of the shift, and is 512 units of the product per hour. **2.** The function $f(x) = x^{2/7} + 1$ decreases at an increasing rate when x is less than 0. When $x = 0, f(x) = 1$ and the function attains an absolute minimum. When x is greater than 0, the function increases at a decreasing rate. **3.** As x increases to $0, f(x)$ increases but does so at a decreasing rate. At $x = 0, f(x) = 4$, and f attains a relative maximum. As x increases from 0 to $6, f(x)$ decreases at an increasing rate. When $x = 6, f(x) = -140$, and the function experiences a point of inflection. As x increases from 6 to $12, f(x)$ decreases at a decreasing rate. When $x = 12, f(x) = -284$, and f attains a relative minimum. When x is greater than 12, $f(x)$ is increasing at an increasing rate. **4.** When x is very small (far to the left of 0), $f(x)$ is very near, but always less than 1. As x increases to $3/7, f(x)$ decreases at an increasing rate. At $x = 3/7, f(x)$ is undefined. In fact, $x = 3/7$ is a vertical asymptote. When x is just beyond $3/7, f(x)$ is unboundedly large. As x increases farther to the right of $3/7, f(x)$ decreases at a decreasing rate and approaches, but is always greater than 1. **5.** Accept the first claim; reject the second claim. **6.** $(-1, 6)$ is a local max; $\left(\frac{7}{9}, -2.43\right)$ is a local min.

Section 3.3

1. Length: 140 ft, Width: 70 ft **2.** The concentration is maximum at $\frac{3}{2800}$ mg/cm^3 seven hours after the drug was administered. **3.** The profit is maximum at \$720 when the price is $\$150 - \$5 \cdot 2 = \$140$. **4.** The volume is maximum at $\frac{1024}{27} = 37.9$ cubic inches when cut is $\frac{4}{3}$ inches by $\frac{4}{3}$ inches. **5.** Accept the claim.

Section 3.4

1. $R(85) = 154{,}121$ and $R'(85) = 6{,}330$. When the company sells 85 units of the product, it realizes a revenue of approximately \$154,121. If the company increases the number of units it sells from 85 to 86, its revenue will increase by approximately \$6,330. Or, the 86th unit will bring in another \$6,330 in revenue. **2.** $C(350) = 4{,}053$ and $C'(350) = 35.38$. The cost of shipping 350 units of the product is approximately \$4,053 and the cost of shipping the 351st unit is \$35.38. Or, if the number of units shipped increases by 1, from 350 to 351, the shipping cost will increase by approximately \$35.38. **3.** $P(40) = 1{,}706$ and $P'(40) = 22.80$. The profit realized by the company on the sale of 40 units of its product is \$1,706. If the company increases the number of sales by 1, from 40 to 41 units, the profit will increase by about \$22.80. Or, the profit realized on the sale of the 41st unit will be about \$22.80. **4.** The maximum profit is \$9,000 and occurs when the level of production is 50 units per month.

Matched Problems
Applied Calculus
Answers

5. The company should place 25 orders of size 40. The minimum cost will be $12,000.

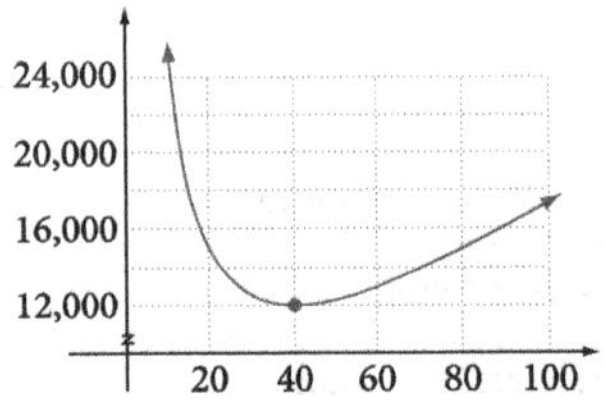

6. A lot size of 40 units per order will minimize inventory cost. **7.** $\varepsilon = 0.67$. If the price of each unit of this product is raised by 1% from $64 to $64.64, the demand for the product will decrease by approximately 0.67% from 300 units to approximately 298 units. **8. a.** Demand is inelastic. The revenue with sales of 300 units at $64 per unit is $19,2000. The revenue with sales of 298 units at $64.64 units is $19,262.72. **b.** The state should not adopt the 1% increase in tuition as $\varepsilon \approx 1.3$ indicates that demand is elastic and that revenue will fall with the increase in the price of tuition.

Chapter 4
Section 4.1

1. a.

x	3^x	$\left(\frac{1}{3}\right)^x$
−3	$0.\overline{037}$	27
−2	$0.\overline{111}$	9
−1	$0.\overline{333}$	3
0	1	1
1	3	$0.\overline{333}$
2	9	$0.\overline{111}$
3	27	$0.\overline{037}$

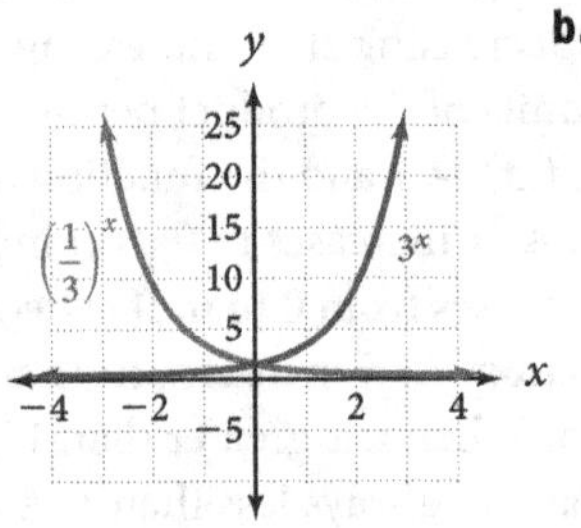

b.

x	$\left(\frac{3}{2}\right)^x$	$\left(\frac{2}{3}\right)^x$
−3	$0.\overline{296}$	3.375
−2	$0.\overline{444}$	2.25
−1	$0.\overline{666}$	1.5
0	1	1
1	1.5	$0.\overline{666}$
2	2.25	$0.\overline{444}$
3	3.375	$0.\overline{296}$

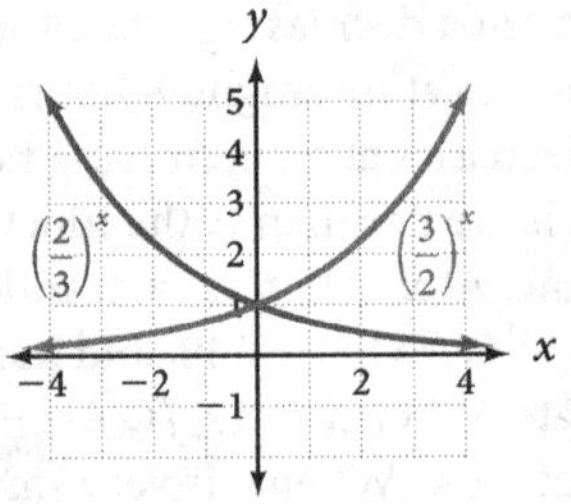

2.

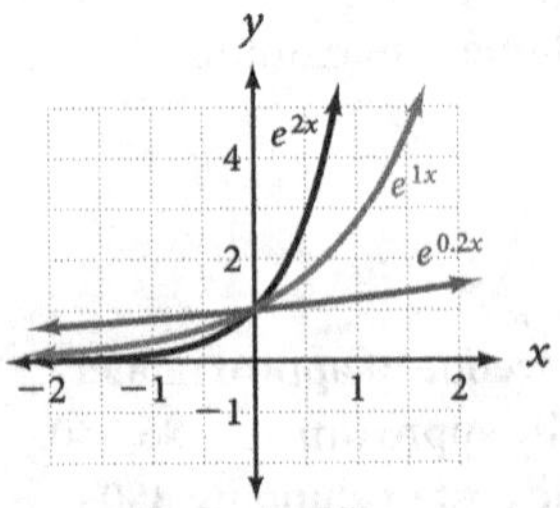

3.

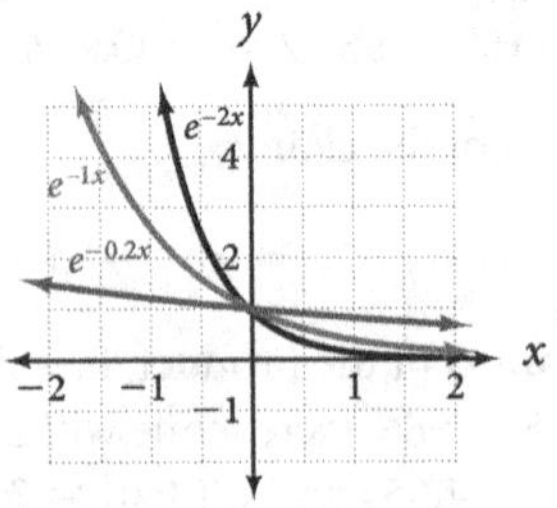

4. About 2,101 bacteria. **5.** The fourth-ranked person participated approximately 15 times. **6.** $5,039.01 **7. a.** About 21,739 cars **b.** About 133,997 cars **c.** At 150,000 cars being sold.

Section 4.2

1. a. $\ln y = 10x$ **b.** $\ln z = 6x - 6$ **c.** $\ln 450 = -0.05x$ **2. a.** $e^{10} = 5x$ **b.** $e^{x+1} = y + 4$ **c.** $e^{-0.005} = x - 10$ **3.** x **4. a.** $\ln 5 + \ln x \approx 1.6094 + \ln x$ **b.** $\ln 3 + \ln x - \ln(x + 9) \approx 1.0986 + \ln x - \ln(x + 9)$ **c.** $4 \ln x$ **d.** $450 \ln x + 1{,}050 \ln y$ **5. a.** $\ln(ab^9)$ **b.** $\ln\left(\frac{dg}{h}\right)$ **6.** $x \approx -0.0602$ **7.** 14 months from March, 2003. **8.** 6 years.

Matched Problems
Applied Calculus
Answers

Section 4.3

1. a. $f'(x) = 5x^3(4\ln x + 1)$ **b.** $y = 502.1x - 1092.0$ **2.** $f'(x) = \dfrac{18x^2 - 7}{6x^3 - 7x + 2}$ **3.** $f'(x) = \dfrac{20}{5x + 6}$ **4.** $f'(x) = \dfrac{10[\ln(4x^2)]^4}{x}$
5. $f'(x) = \dfrac{-24}{x[\ln(6x^3)]^2}$ **6.** $y' = \dfrac{-2xy}{x^2 + 2y}$

Section 4.4

1. $f'(x) = 7x^3e^x(4 + x)$ **2. a.** $y = 5x$ **b.** $y = ax$ **3. a.** $f'(x) = (12x^2 - 7)e^{4x^3 - 7x}$ **b.** $f'(x) = 2(4x^3 - 7x)(12x^2 - 7)e^{(4x^3 - 7x)}$
4. $f'(2) = 0.4584$ **5.** $f''(x) = 36x^2e^{3x^4 - 8}(1 + 4x^4)$ **6.** $f'(x) = -5e^{-2x}\left(2\ln 6x - \dfrac{1}{x}\right)$ **7.** $f'(0) = \dfrac{3}{8}$
8. a. $9e^{9x}$ **b.** $-3e^{-3x}$ **c.** $-0.01e^{-0.01x}$ **d.** $-5e^{-0.5x}$ **9. a.** -29% **b.** $R'(1) = -81{,}373.66$ **c.** $R'(5) = -25{,}509.52$
10. a. 3.4 mg **b.** $t \approx 44$ hours **c.** $A'(5) = -0.073$ mg/hr

Chapter 5
Section 5.1

1. a. $7x + C$ **b.** $r + C$ **2. a.** $\dfrac{x^7}{7} + C$ **b.** $\dfrac{4}{9}x^{9/4} + C$ **c.** $\dfrac{-1}{2x^2} + C$ **3. a.** $2x^5 + C$ **b.** $\dfrac{7x^4}{4} + C$
4. a. $\dfrac{5x^4}{4} - 4x^2 + 4x + C$ **b.** $\dfrac{20x^{6/5}}{3} - 3x + C$ **5.** $4e^x + x^3 + C$
6. a. $5\ln|x| + C$ **b.** $6\ln x - 2e^x - 5x + C$ **c.** $\dfrac{3x^2}{2} - 8x + 4\ln x + C$ **7.** $f(x) = x^4 + 3x^3 + x$
8. a. $C(x) = -0.01x^2 + 51x + 4040$ **b.** \$18,440

Section 5.2

1. $\ln|9x^2 + 8x - 3| + C$ **2. a.** $\dfrac{1}{2}\ln|3x^2 + 2x| + C$ **b.** $4\ln|3x^2 + 2x + 5| + C$ **3. a.** $\dfrac{4}{5}e^{5x-8} + C$ **b.** $2x^2 + \dfrac{e^{-2x}}{2} + C$
4. a. $e^{6x^2+1} + C$ **b.** $2e^{4x^2+5x} + C$ **5.** $\dfrac{(5x^2 - 7x)^5}{5} + C$ **6.** $3(2x^2 + x)^{2/3} + C$ **7.** $\dfrac{3(x-5)^{7/3}}{7} + \dfrac{15(x-5)^{4/3}}{4} + C$
8. a. $\dfrac{e^{4x}}{4} + C.$ **b.** $\dfrac{5e^{\frac{3}{5}x}}{3} + C$

Section 5.3

1. a. $\dfrac{33}{2}$ **b.** 128 **2. a.** $5\ln(3) \approx 5.4931$ **b.** $3\ln(3) \approx 3.2958$ **3.** $5\ln(3) \approx 5.4931$

Section 5.4

1. 102.89 **2.** 120.6 square units **3.** $-\dfrac{32}{3}$ **4.** $\dfrac{27}{2} + \dfrac{17}{6} = \dfrac{49}{3}$ square units

Section 5.5

1. 2 **2.** 1 **3.** Does not exist, or diverges **4.** The integral does not converge **5.** $t = 20{,}000$ kg

6. Bioavailability $= F = \dfrac{\int_0^\infty 100(e^{-0.4t} - e^{-1.0t})\,dt}{\int_0^\infty 80e^{-0.4t}\,dt} = 0.75$ which means that if the dose is taken orally it will be only about 75% as effective as a dose taken intravenously.

Matched Problems
Applied Calculus
Answers

Section 5.6

1. $\frac{1}{4}x^4 \ln x - \frac{1}{16}x^4 + c = \frac{1}{16}x^4(4\ln x - 1) + C$ **2.** $\frac{1}{3}xe^{3x} - \frac{1}{9}e^{3x} + C = \frac{1}{9}e^{3x}(3x - 1) + C$

3. $\frac{x^3 \ln x}{3} - \frac{x^3}{9} + C = \frac{x^3}{9}(3x - 1) + C$ **4.** $N(t) = -300te^{-0.7t} - 428.571e^{-0.7t} + 428.571$

Chapter 6
Section 6.1

1. $A = \int_1^6 [(-3x + 26) - (x^2 - 10x + 32)]\,dx = \frac{125}{6}$ **2.** $A = \int_4^6 [(-2x^2 + 12x + 15) - (2x^2 - 28x + 99)]\,dx = \frac{88}{3}$

3. $A = \int_{-5}^4 [(-x^2 + 8x + 70) - (x^2 + 10x + 30)]\,dx = 243$

4. $A = \int_{-3}^0 [(x^3 - 5x + 20) - (4x + 20)]\,dx + \int_0^3 [(4x + 20) - (x^3 - 5x + 20)]\,dx = \frac{81}{2}$

5. Difference $= \int_0^{10} [(80e^{0.10x}) - (80e^{0.08x})]\,dx = 149$ million dollars

Section 6.2

1. $CS = \int_0^{15} [D(x) - p_e]\,dt = \int_0^{15} [1200 - 4x^2 - 300]\,dt = 9000 \rightarrow \$9,000$

Interpretation
The consumer's surplus is the total difference between the price that consumers are willing to pay for the product and the price that they actually do pay for the product, that is, the market price of the product. The \$9000 is the total amount of money saved by consumers of the product when they buy each unit for the market price of \$300 rather than at a higher price they would be willing to pay.

2. $CS = \int_0^{x_e} [D(x) - p_e]\,dx = \int_0^{56.69} [450 - 0.09x^2 - 160.71]\,dx = 10,934.22 \rightarrow \$10,934.22$

$PS = \int_0^{x_e} [p_e - S(x)]\,dx = \int_0^{56.69} [160.71 - 0.05x^2]\,dx = \$6,074.19 \rightarrow \$6,074.19$

Interpretation
The consumer's surplus is the total difference between the price that consumers are willing to pay for the product and the price that they actually do pay for the product, that is, the market price of the product. The \$10,934.22 is the total amount of money saved by consumers of the product when they buy each unit for the market price of \$160.71 rather than at a higher price they would be willing to pay.

The producer's surplus is the total difference between the price at which producers are willing to supply the product and the price they actually receive for the product, that is, the market price. The \$6,074.19 is the total amount of money realized by producers of the product when they sell each unit for the market price of \$160.71 rather than at a lower price they would be willing to accept.

Matched Problems
Applied Calculus
Answers

3. $CS = \int_0^{x_e} [D(x) - p_e]\, dx = \int_0^{149.59} [-0.002x^2 + 174 - 129.25]\, dx = 4{,}462.55 \rightarrow \$4{,}462.55$

$PS = \int_0^{x_e} [p_e - S(x)]\, dx = \int_0^{149.59} [129.25 - (0.0006x^2 + 0.44x + 50]\, dx = 6{,}262.55 \rightarrow \$6{,}262.55$

Interpretation
The $4,462.55 is the total amount of money saved by consumers of the product when they buy each unit for the market price of $149.59 rather than at a higher price they would be willing to pay.

The $6,262.55 is the total amount of money realized by producers of the product when they sell each unit for the market price of $149.59 rather than at a lower price they would be willing to accept.

Section 6.3

1. $A = \int_0^4 6{,}350{,}000\, dt = 6{,}350{,}000t \Big|_0^4 = \$25{,}400{,}000$

Interpretation
If revenue flows into the account of the chain retail operation at the constant rate of $f(t) = 6{,}350{,}000$ dollars per year, and if as it is received, it is deposited into a non-interest bearing account, the amount of money that will accumulate in the account over a 4-year period will be $25,400,000.

2. $A = Pe^{rt} = \$320{,}000e^{(0.04)(5)} \approx \$390{,}848.88$

Interpretation
The $320,000 is the amount of money that would have to be invested in an account paying 4% compounded continuously now to get about $390,849 five years from now.

3. $A = \int_0^n Pe^{rt}\, dt = \int_0^{120} \$1{,}000e^{0.00417t} \approx \$155{,}693.10$

Interpretation
The future value of $1000 invested at the end of each month for 10 years (120 months) into an annuity account paying 5% compounded continuously is approximately $155,693.10. That is, the $120,000 investment is actually worth about $155,693.10.

4. a. $A = \int_0^{12} 1{,}000\, e^{0.00417t} \approx \$12{,}305.06$ **b.** $A = \int_0^{240} 1{,}000\, e^{0.00417t} \approx \$412{,}387.64$ **5.** $A = Pe^{-rt} = \$30{,}000e^{-0.03 \cdot 10} \approx \$22{,}224.55$

Section 6.4

1. $\frac{1}{y^3 + 5}\, dy = 3x^3 dx$ **2.** $y = \frac{5}{2}\ln(x^2 - 5) + C$ **3.** $y = \frac{1}{x^4 - 15}$

4. $y = \sqrt[3]{4 + 4x^3}$ **5.** $y = e \cdot e^{\frac{3x^5}{5}}$ or $y = e^{\frac{3x^5}{5} + 1}$ **6.** $y = 12e^{-2x}$

7. $N(t) = \frac{4(t + 25)^2}{5}$. Twelve months after the start of the campaign, the company expects to have about 1,095 subscribers.

Matched Problems
Applied Calculus
Answers

Section 6.5

1. On January 1, 2015, if this model holds, the property will appraise at about \$442,975. This estimate uses the model $P(t) = 840{,}000e^{-0.106648t}$ **2.** $P(t) = 90 + 633e^{-0.005335408t}$ After 5 hours (300 minutes), the temperature of the steel bar is about 21°C. **3. a.** $P(t) = 370{,}000 - 369{,}880e^{-0.0045833t}$ **b.** About 38,649 people in the network will have heard the information 24 hours after its release. **c.** About 1,753 hours after its release, all the members of the network will know about the information. **4. a.** $P(t) = \dfrac{5{,}740{,}000}{410 + 13{,}590e^{-0.03937t}}$ **b.** 7,969 plants **c.** 90% of the plants will have the disease in about 145 – 36 months = 109 months from now.

Section 6.6

1. Yes, both conditions are satisfied. **2.** 0.1975 **3.** $P(10 \le X \le 15) = \int_{10}^{15} \frac{1}{15}\,dx = \frac{1}{3}$ If you arrive randomly at this bus stop often, for about $\frac{1}{3}$ of your arrivals you would have to wait less than 5 minutes.

4. $P(X \le 75) = \int_0^{75} 0.01e^{-0.01x}\,dx \approx 0.5276$ The manufacturer can expect that approximately 52.76% of its new wave-shapers will last at most 75 months. **5.** $P(5.0 \le X \le 5.7) = 0.2515$

Chapter 7
Section 7.1

1. \$10,075,000

> **Interpretation**
> If the Ventus Potestatem wind energy company manufactures and sells thirty of the 5 kW turbines, forty of the 35 kW turbines, and 20 of the 50 kW turbines, it will realize a revenue of \$10,075,000.

2. a. 210,600 units

> **Interpretation**
> If the company uses 243 units of labor and 1024 units of capital, its level of production will be 210,600 units.

b. 285 units

> **Interpretation**
> If the company decreases the number of units of labor it uses by 1, from 243 to 242, and increases the number of units of capital is uses by 10, from 1024 to 1034, the level of production will decrease by about 285 units.

Matched Problems
Applied Calculus
Answers

Section 7.2

1. a.

> **Interpretation**
> If, when x is 5 and y is 2, x is increased by 1 unit, from 5 to 6 and the value of y remains constant at 2, the function will increase by about 47 units.

b.

> **Interpretation**
> If, when x is 1 and y is 2, x is increased by 1 unit, from 1 to 2 and the value of y remains constant at 2, the function will decrease by about 17 units.

2. a. $\frac{\partial f}{\partial x} = 20x^3 + 4$ **b.** $\frac{\partial f}{\partial y} = -6y^2 + 1$ **c.** $\frac{\partial f}{\partial x} f(2, 3) = 164$; If, when x is 2 and y is 3, x is increased by 1 unit, from 2 to 3 and the value of y remains constant at 3, the function will increase by about 164 units. **d.** $\frac{\partial f}{\partial y} f(2, 3) = -53$; If, when x is 2 and y is 3, y is increased by 1 unit, from 3 to 4 and the value of x remains constant at 2, the function will decrease by about 53 units.

3. a. $f_x(x, y) = 3x^2 + 9x^2y^2$ **b.** $f_y(x, y) = 6x^3y - 12y$ **c.** $f_x(4, 1) = 192$ If, when x is 4 and y is 1, x is increased by 1 unit, from 4 to 5 and the value of y remains constant at 1, the function will increase by about 192 units. **d.** $f_y(4, 1) = 372$ If, when x is 4 and y is 1, y is increased by 1 unit, from 1 to 2 and the value of x remains constant at 4, the function will increase by about 372 units. **4.** $f_x = 6(2x^5 - 7y^3)^5(10x^4) = 60x^4(2x^5 - 7y^3)^5$ **b.** $f_y = 6(2x^5 - 7y^3)^5(-21y^2) = -126y^2(2x^5 - 7y^3)^5$

5. a. $f_x = \frac{8}{x}$ **b.** $f_y = 2e^{2y+5z}$ **c.** $f_z = 5e^{2y+5z} + \frac{8}{z}$ **6.** (0, 0)

Section 7.3

1. (3, 1)

> **Interpretation**
> The critical point (3, 1) may produce a relative maximum, a relative minimum or a saddle point.

2. (57, −82)

> **Interpretation**
> The critical point (57, −82) may produce a relative maximum, a relative minimum or a saddle point.

3. (1, 0), (1, −2), (1, 2)

> **Interpretation**
> The critical points (1, 0), (1, −2), (1, 2) may produce some combination of relative maximums, relative minimums or saddle points.

Matched Problems
Applied Calculus
Answers

4. No critical values

> **Interpretation**
> This function has no critical points. Therefore, it has no relative maximums, no relative minimums, and no saddle points.

5. The critical point is $(3, -10)$.

> **Interpretation**
> The critical point $(3, -10)$ produces the relative minimum point $(3, -10, -119)$.

6. The critical values are $(3, 0)$, $(3, 1)$, $(-3, 0)$, $(-3, 1)$.

$f(a, b)$	Conclusion
$f(-3, 0) = 63$	Rel. max at $(-3, 0, 63)$
$f(-3, 1) \approx 62.33$	Saddle point at about $(-3, 1, 62.33)$
$f(3, 0) = -45$	Saddle point at $(3, 0, 45)$
$f(3, 1) = -45.67$	Rel. min at about $(3, 1, -45.67)$

7. $x = 1{,}375$ and $y = 675$

> **Interpretation**
> To maximize its profit, Drut should manufacture and sell 1,375 units of wire in Maryland and 675 units in Georgia.

Section 7.4

1. The critical point is $(-15, -4)$ and it produces the constrained relative minimum $(-15, -4, 1)$.
2. The critical point is $(4, -2)$ and it produces the constrained relative minimum $(4, -2, 100)$.
3. The critical value $(66.67, 6{,}000{,}000)$ produces the maximum point $(66.67, 6{,}000{,}000, 1{,}877{,}481)$.

> **Interpretation**
> To maximize production, the manufacturer should allocate 66.67 hours of labor each working day of the year and \$6,000,000 of capital.

Section 7.5

1. $df(6{,}561, 256) = -794.20$

> **Interpretation**
> If the number of units of labor is increased from 6,561 units to 6,566 units and capital is decreased from 256 units to 250 unit, the number of units of output will decrease by about 794.

Matched Problems
Applied Calculus
Answers

Section 7.6

1. $\int 12xy^3\, dy = 3xy^4 + h(x)$, where $h(x)$ represents the constant of integration, a function of x in this case.

2. $V = 619.5$ cubic units **3.** $V = \frac{7}{60} \approx 0.117$ cubic units **4.** $V = \frac{7}{60} \approx 0.117$ cubic units

5. Let X represents the measure of a diameter, and Y represents the measure of a thickness. Then

$$P(0.303 \le X \le 0.307 \text{ and } 12.49 \le Y \le 12.51) = \int_{0.303}^{0.307}\int_{12.49}^{12.51} 1{,}000\, dydx = 0.08$$

Interpretation
The probability that a randomly quartz crystal produced by this company will have a diameter between 12.49 and 12.51 mm and a thickness between 0.303 and 0.307 mm is 0.08. That is, the manufacturer can expect about 8% of the quartz crystals that are produced will have diameters that differ from the expected diameter by less than 0.01 mm and thicknesses that differ from its expected thickness by less than 0.002 mm.

6. $\iint_R P(x, y)\, dA = \int_0^{20}\int_0^{10} 300e^{-0.2x-0.3y}\, dydx = 4{,}664.05$

Interpretation
There are about 4664 million molds on the surface.

Section 7.7

1. The average number is 68,109. $\frac{1}{(365-0)}\int_0^{365} 10{,}000te^{-0.02t^2}\, dt = 684.93$

2. The average monthly profit is $62,667. $\frac{1}{100-0}\cdot\frac{1}{100-0}\int_0^{100}\int_0^{100} [6{,}000 - 2x^2 + 1{,}000x + 4(y-100)^2]\, dydx = \$90{,}000$